SpringerBriefs in Education

We are delighted to announce SpringerBriefs in Education, an innovative product type that combines elements of both journals and books. Briefs present concise summaries of cutting-edge research and practical applications in education. Featuring compact volumes of 50 to 125 pages, the SpringerBriefs in Education allow authors to present their ideas and readers to absorb them with a minimal time investment. Briefs are published as part of Springer's eBook Collection. In addition, Briefs are available for individual print and electronic purchase.

SpringerBriefs in Education cover a broad range of educational fields such as: Science Education, Higher Education, Educational Psychology, Assessment & Evaluation, Language Education, Mathematics Education, Educational Technology, Medical Education and Educational Policy.

SpringerBriefs typically offer an outlet for:

- An introduction to a (sub)field in education summarizing and giving an overview of theories, issues, core concepts and/or key literature in a particular field
- A timely report of state-of-the art analytical techniques and instruments in the field of educational research
- A presentation of core educational concepts
- An overview of a testing and evaluation method
- A snapshot of a hot or emerging topic or policy change
- An in-depth case study
- A literature review
- A report/review study of a survey
- An elaborated thesis

Both solicited and unsolicited manuscripts are considered for publication in the SpringerBriefs in Education series. Potential authors are warmly invited to complete and submit the Briefs Author Proposal form. All projects will be submitted to editorial review by editorial advisors.

SpringerBriefs are characterized by expedited production schedules with the aim for publication 8 to 12 weeks after acceptance and fast, global electronic dissemination through our online platform SpringerLink. The standard concise author contracts guarantee that:

- an individual ISBN is assigned to each manuscript
- each manuscript is copyrighted in the name of the author
- the author retains the right to post the pre-publication version on his/her website or that of his/her institution

Nicola Acocella

Understanding Education in the European Union

From Different Systems to an Ideal One

Springer

Nicola Acocella
Sapienza University of Rome
Rome, Italy

ISSN 2211-1921 ISSN 2211-193X (electronic)
SpringerBriefs in Education
ISBN 978-3-031-62064-5 ISBN 978-3-031-62065-2 (eBook)
https://doi.org/10.1007/978-3-031-62065-2

This Springer imprint is published by the registered company Springer Nature Switzerland AG
The registered company address is: Gewerbestrasse 11, 6330 Cham, Switzerland

Education is the most powerful weapon which you can use to change the world

—Nelson Mandela said in his speech at Madison Park High School, Boston, 23 June 1990

Introduction

JEL classifications: I20; I260

Definition

Education is the process of understanding of, or information about, a subject that a person can get by experience or study, with the help of educators.

From Wikipedia, Education, https://en.wikipedia.org/wiki/Education, we learn that the precise definition of education is the object of many debates, as to: its aims; whether and why the change in the person induced by education is an improvement; the need to distinguish its role of critical thinking from indoctrination.

Education has been defined as an activity aimed at transmitting or acquiring knowledge, skills and the character traits of a person or as an activity tending to develop the powers of reasoning and judgment forming and enriching the spirit and the process of socialisation of a person, in particular directing students toward their eventual role in the society.

The term education derives from the Latin words educare, meaning 'bring up, rear, educate', primarily related to the mind, and educere, meaning 'bring out, lead forth' (Wikipedia).

There are many debates about its precise definition, for example, about which aims it tries to achieve. A further issue is whether part of the meaning of education is that the change in the student is an improvement. Some researchers stress the role of critical thinking to distinguish education from indoctrination. These disagreements affect how to identify, measure and improve forms of education. The term can also refer to the mental states and qualities of educated people. By the term 'education' we refer not only to the transmission of the values and accumulated knowledge of a society, but also to the discipline that is concerned with the methods of teaching and learning.

As noted by the World Bank, 'Education is a human right, a powerful driver of development, and one of the strongest instruments for reducing poverty and

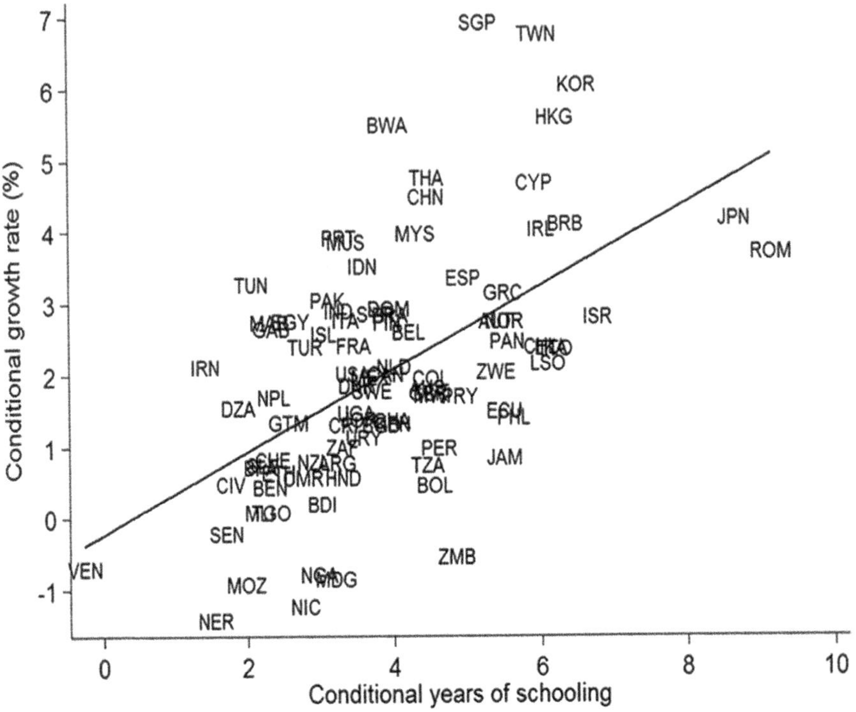

Fig. 1 Years of schooling and economic growth rates, %, *Source* Hanusek, Woessmann (2015) *Note* The figure plots the average annual rate of growth in per capita GDP over the years 1960–2000 against years of schooling at the beginning of the period for a sample of 92 countries. Both growth and education are expressed as conditional on the initial level of per capita GDP to account for a significant conditional convergence effect.

improving health, gender equality, peace, and stability. It delivers large, consistent returns in terms of income, and is the most important factor to ensure equity and inclusion' (World Bank, 2023, https://www.worldbank.org/en/topic/education/overview).

From the point of view of individuals, education promotes employment, earnings and health and reduces poverty. In particular, at a world level, it has been shown that hourly earnings increase by 9% for every extra year of schooling. For societies, education increases long-term economic growth, spurs innovation, strengthens institutions and promotes social cohesion.

For the relationship between education and growth we can refer to Fig. 1. This shows that a higher number of schooling years adds to economic growth.

Investment in people's education can reduce extreme poverty, by developing human capital.

The pandemic has, however, aggravated the state of ignorance, leading to the worst crisis in education of the last century. In fact, at the world level between February 2020 and February 2022, education systems were fully closed for 141 days on average—reaching a peak of 273 days in South Asia and 225 days in Latin America and the Caribbean—which could cause an increase in the share of children living in 'Learning Poverty' from 57% to 70%. There were negative effects of school closures beyond learning. In fact, children affected by the pandemic could lose a $21 trillion in their lifetime earnings, around 17% of today's global GDP. Apart from total negative effects, also inequalities derived from the COVID-19 rose. They hit children from disadvantaged households—who had difficulties in benefitting from remote learning, due to a lack of electricity, connectivity, etc.—who were 463 million.

In low- and middle-income countries, given the long school closures and the wide digital divide that hindered the effectiveness of remote learning during school closures, putting in jeopardy the SDG4 targets—i.e. the goals established in 'Sustainable Development Goal 4 (SDG4) of the 2030 Agenda for Sustainable Development', one billion children saw their personal education interrupted for more than a year. For many, the interruption was for 2 years.

Children and youth in most countries suffered major learning losses during the pandemic. Rigorous empirical evidence from various countries, including low-, middle- and high-income contexts across regions, reveals very steep losses. School closures and ineffective remote learning caused students to miss out on learning and to also forget what they had learned: on average, for every 30 days of school closures, students lost about 32 days of learning.

The staggering effects of school closures reached beyond learning. This generation of children could lose a combined total of US$ 21 trillion in lifetime earnings in present value or the equivalent of 17% of today's global GDP—a sharp rise from the 2021 estimate of a US$ 17 trillion loss.

COVID-19 created an inequality catastrophe. Almost all countries provided some form of remote education during school closures, but there was high inequality in access and uptake between and within countries. Children from disadvantaged households were less likely to benefit from remote learning than their peers, often due to a lack of electricity, connectivity devices and caregiver support. Girls, students with disabilities and the youngest children also faced significant barriers to engaging in remote learning. Overall, at least a third of the world's schoolchildren—463 million globally—were unable to access remote learning during school closures.

In addition, both children's mental health and children with disabilities have been negatively affected by the pandemic, due to the ensuing isolation. School disruptions affected in particular the youngest children. Various measures were implemented to reduce the negative effects of the pandemic. Apart from the closure of early childhood education, with no support for remote learning, these negative effects were: individualised self-learning programmes (computer-assisted or pencil-and-paper based), automatic re-enrollment of students in school, mobilisation campaigns to bring students back to school, early warning systems to identify students at risk of dropping out, tutoring programmes (in person or remote) or financial support for tutoring, psychosocial and mental health support to students (OECD, 2022).

Formal education takes place in institutional frameworks of the kind of public schools. There are various levels of education: early childhood education, primary education, secondary education and tertiary education.

In addition to augmenting knowledge, education has an important role in socializing children into society. It also stimulates economic growth and 'raises awareness of local and global problems'.

Many factors—such as motivation, intelligence, personality of both teachers and students, their socioeconomic status, ethnicity, gender, educational technology, teacher quality and parent involvement—influence the success of education. The invention of writing, first, and the invention of the printing press, later, caused a shift from informal to formal education and its diffusion among the different groups of the population, being initially limited to elites and religious groups. In the 18th and 19th centuries, public primary education led to its further diffusion, particularly in Western countries. Diffusion in other countries is more recent, as we will see below.

With growth of societies, which implies their higher complexity, education needs to become more specialised and imparted by several teachers, especially as students grow up.

Types of Education

There are many types of education, going from formal education—which takes place in a complex institutional framework, like public schools—to informal education, outside the schooling system. Both are divided into the different levels already seen. Education socialises children into society, and there are various organisations—also international, such as UNESCO—that promote it for all children. Its success depends on psychological factors (such as motivation, intelligence and personality), social factors (socioeconomic status, ethnicity and gender) and other factors that include educational technology, teacher quality and parent involvement.

References

OECD. (2022). *Education at a glance 2022: OECD indicators*. OECD Publishing. https://doi.org/10.1787/3197152b-en

World Bank. (2023). Education April 11. https://www.worldbank.org/en/topic/education/overview

Contents

List of Figures

List of Tables

Chapter 1
Education Models

Introduction

This chapter deals with:

(i) Organisational models (Sect. 1.1). These can be referred to: The compulsory or voluntary attendance of the various grades of school (Sect. 1.1.1); the public or private provider of the education in various countries and with specific reference to Europe and the OECD countries (Sect. 1.1.2); the issues dealt with in these two sub-sections are strictly tied one to the other, since public provision is usually associated to compulsory education;

(ii) Models of education (Sect. 1.2), which include: models of financing (Sect. 1.2.1), a subject that is tied to that of organisational model, and methods and models of teaching and learning (Sect. 1.2.2), a topic which is indeed independent from the previous ones.

1.1 Organisational Models

1.1.1 Compulsory Education

Compulsory education was introduced since the eighteenth century in Denmark and Prussia and is now a reality practically all over the world, except in a few countries, Bhutan, Papua New Guinea, Solomon Islands and the Vatican City. Usually, it should last for around 8 years, starting from the age of about 5.

Introducing compulsory schooling tends to increase the level of education, since there are some young people belonging to marginalised groups who would tend to leave school early or neglect attending it completely. This innovation is certainly expensive for the families and for state or other public bodies, but there are economic returns for those that attend schools and for the system as a whole. There are also social benefits, such as: increases in the ability to communicate, use to teamwork,

N. Acocella, *Understanding Education in the European Union*,
SpringerBriefs in Education, https://doi.org/10.1007/978-3-031-62065-2_1

ability to develop personal talents, increase of tolerance and respect for authority, 'crime reduction, improvements in financial literacy, and improvements in mental cognition in older age' or health (Harmon, 2017), as well as increases in social capital, i.e., in all the aspects of social life that allow people to operate more effectively together in order to pursue common targets. An insufficient number of teachers, possibly due to scarcity of resources, and poorly designed curricula may prevent or reduce positive effects.

1.1.2 State and Private Schools

Schools may be financed and run by the state—or other public bodies, such as municipalities and provinces—or by privates, such as churches. Sometimes they are financed by public bodies, but run by individuals or corporations.

In the medieval period, apart from public universities that were established in the thirteenth century, such as the University of Bologna, the University of Paris, and Oxford University, education was mainly provided or guided by religious or other private authorities or tutors. The increased availability of written media due to the invention and popularisation of the printing press in the fifteenth century had a significant influence on the literacy level of the population. In the eighteenth and nineteenth centuries public education for all was facilitated by these changes and also intergovernmental organisations (see, e.g., the Universal Declaration of Human Rights, the Convention on the Rights of the Child, the Education for All initiative, the Millennium Development Goals, and the Sustainable Development Goals). Therefore, the literacy rate rose. In 2023 this rate for all males and females that are at least 15 years old is 86.3%, reaching in the last few years a level of 99.99% in Uzbekistan and 99.97% in Ukraine.

According to the evaluations of PISA—which stands for the acronym of the OECD *Programme for International Student Assessment*[1]—in 2018 on average in the OECD countries only 2 out of 10 students attended a private school, having increased a little since 2010. The percentage was however higher than 50% in Chile, Hong Kong (China), Lebanon, Macao (China), the Netherlands, the United Arab Emirates and the United Kingdom.

A good summary view of the evolution of education in Europe is offered by Porcher (2020).[2] He notes that the state initially began to be involved in schooling during the first half of the nineteenth century, to oversee and promote its development in Europe under a double impulse: the spread of the revolutionary ideals of the French revolution and the needs of industrialisation, sometimes in opposition to the

[1] PISA is a worldwide study in member and non-member nations tending to evaluate educational systems by measuring pupils' scholastic performance. It was first introduced in 2000 and the last tests took place in 2018 in 79 countries. PISA tests are used in order to advice education national and international policymakers, as they can offer insights about the factors behind the different performances within and between countries.

[2] For a longer history of education see Chap. 3 of the present book.

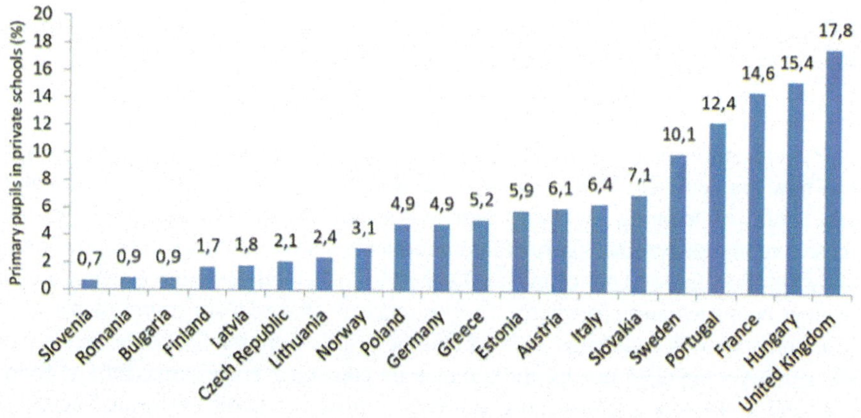

Fig. 1.1 Percentage of primary school pupils attending private schools in European countries where school ownership belongs prevailingly to the public sector, 2015. *Source* Radó (2019)

Church, even if for some time there was a social differentiation (classical humanities for high-ranked people and basic primary education, for ordinary people). In most European countries, primary education became compulsory, even if in some countries, such as Italy, the obligation was disattended, either for ignorance or the needs of children to do homework or work in the fields. In many cases education was completely free, while in others—such as in Italy, England and Denmark—it was partially free. During the twentieth century, in some countries such as Spain and the United Kingdom it was partially private. A single curriculum was introduced for elementary and lower-secondary schools, to promote mass spread of education, in general after WWII, while in the United States that had already happened in the early twentieth century. More recently (see Sect. 5.5), also higher education has become common. While the European Union encourages cooperation, states remain sovereign, with the distinctive features of various educational traditions remaining quite strong.

A recent picture of the current situation in Europe is offered by Fig. 1.1.

The majority of school funding comes from the central government level, but other actors, including lower-level governments, also contribute to raising funds. Private operators' funding and international sources have increased considerably in recent years in some countries (OECD, 2017).

As to attendance at private schools,[3] in 2018 on average across OECD countries, only 20% of students did so, but in some countries or places—Chile, Hong Kong, Lebanon, Macao, the Netherlands, the United Arab Emirates and the United Kingdom—the percentage rose to 50%. These percentages remained stable since

[3] Private or independent schools—either elementary, middle and secondary schools or even colleges and universities—are schools privately owned and funded without—or scarce—assistance of public governments.

2000, on average across OECD countries, but increased in 10–12 education systems and decreased in 2–8, depending on the years.

On average across OECD, students in public schools scored higher than students in private schools, in reading, mathematics and science.

PISA examines the knowledge of students in reading, mathematics and science, and how they can exploit this knowledge. Results from PISA indicate the quality and equity of learning outcomes attained around the world, thus providing the most comprehensive and rigorous international assessment of those outcomes, and allowing educators and policy makers to learn from the policies and practices applied in their countries and elsewhere. The last survey—the seventh round of triennial assessments—was conducted in 2018 and the results were published on September 29, 2020. Among other results, the survey found that on average across OECD countries and in 40 education systems, students in private schools performed better in reading than students in public schools, if students' and schools' socio-economic profile is not taken into account (see Table 1.1). When this is done, the reverse is true. In general, then, the preparation offered by public schools was higher, in particular if account is taken of the students' socio-economic profile. After accounting for students' and schools' socio-economic profile, students of public schools led the others by 14 score points in general across OECD countries and—more specifically—in 19 countries. At the system level, after accounting for per capita GDP, the higher the share of students in private-independent schools the lower their mean performance in reading, mathematics and science.

In addition to the socio-economic profile, results depend on other factors, such as the attendance by students of pre-primary schools or the number of students per class and the quality of teachers (OECD, 2022, p. 203).

Across OECD countries the number of students enrolled in private schools remained stable since 2000 on average, but increased in 10 countries and decreased in 2. More recently, after 2015, students in private schools increased in 12 systems and decreased in 8.

On average across OECD countries in 2018, 78% of students attended a school whose principal reported that there is at least one other school in the same area available for students. In most countries, and on average across OECD countries, school availability and perhaps competition is associated with better reading performance, before accounting for socio-economic disparities; but—as said—there is no difference in student performance after accounting for students' and schools' socio-economic profile.

Table 1.1 Different reading ability, state and private schools, 2018. *Source* OECD (2020)

Difference in reading score between students in public schools and students in private schools (private-dependent and private-independent combined

	Percentage of students attending:		
	Government or public schools[1]	Government-dependent private schools[2]	Government-independent private schools[3]
Serbia	96.8	0.0	3.2
Chinese Taipei	68.4	14.0	17.6
Hong Kong (China)	8.6	91.1	0.3
Japan	66.3	3.6	30.1
Portugal	86.6	8.8	4.6
Indonesia	53.5	30.0	16.6
Turkey	87.9	1.1	11.0
Netherlands	36.5	63.4	0.1
Thailand	84.0	8.5	7.4
Latvia	98.5	0.8	0.7
Switzerland	95.5	0.7	3.8
Italy	96.4	1.7	1.9
Czech Republic	93.6	5.8	0.6
Saudi Arabia	86.7	5.8	7.5
France	80.0	11.7	8.3
B-S-J-Z (China)	85.7	0.3	14.0
Hungary	79.4	19.3	1.2
Estonia	96.1	2.3	1.6
United States	93.0	2.2	4.8
Korea	60.6	35.5	3.9
Sweden	80.7	19.2	0.1
Denmark	72.0	21.7	6.3
Mexico	87.9	4.2	7.9
Malaysia	93.8	0.5	5.7
Singapore	90.5	3.0	6.6
United Kingdom	34.0	59.8	6.2
OECD average	81.9	13.2	4.9
Austria	87.6	10.8	1.6
Germany	96.1	3.4	0.6
Jordan	78.9	1.4	19.6
Croatia	97.6	1.5	0.9
Canada	91.8	3.4	4.8
Luxembourg	82.3	15.1	2.6
Morocco	92.7	3.0	4.3
Finland	95.9	4.1	0.0
North Macedonia	98.7	0.6	0.7
Australia	57.6	28.2	14.2
Kazakhstan	91.7	2.3	6.0
Slovak Republic	87.7	11.8	0.5
Chile	34.0	56.2	9.8
Albania	88.8	1.5	9.7
Lebanon	48.4	31.2	20.4
Poland	95.5	3.6	0.9
New Zealand	94.2	0.0	5.8
Georgia	89.3	1.1	9.5
Philippines	82.3	10.6	7.1
Malta	54.9	31.4	13.7
Dominican Republic	83.1	6.0	10.9
Lithuania	95.8	3.0	1.2
Qatar	57.3	1.5	41.2
Greece	94.9	1.4	3.7
Uruguay	84.1	0.5	15.3
Panama	81.9	7.2	10.8
Peru	75.2	0.3	24.6
Slovenia	97.5	2.5	0.0
United Arab Emirates	38.0	23.8	38.2
Argentina	68.4	24.8	6.8
Costa Rica	86.2	0.7	13.1
Colombia	81.4	1.4	17.2
Brunei Darussalam	84.3	3.5	12.1
Macao (China)	5.7	85.3	9.0
Brazil	85.0	4.1	10.9

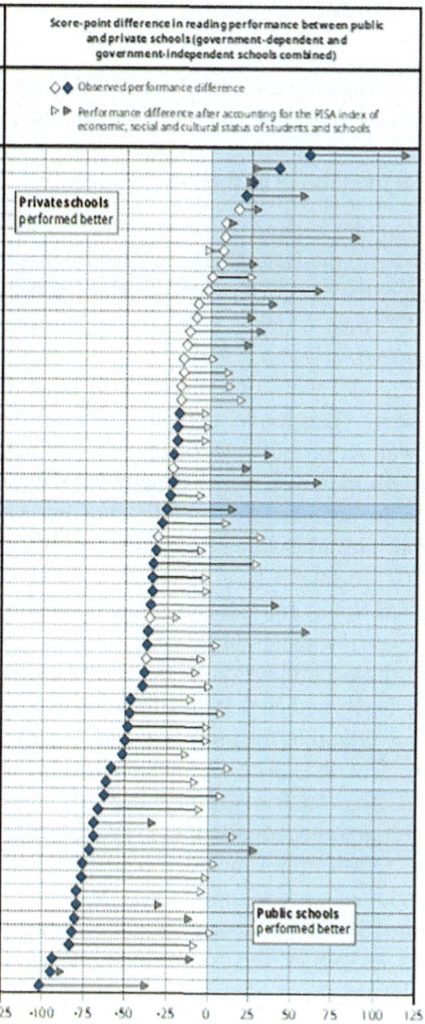

Source OECD (2020)

1.2 Models of Education

1.2.1 *Models of Financing*

There are at least five different models of financing education. Some of them, at least, are typically applied in the countries and the regions we will deal with in Chaps. 4, 6, and 7.

These models are as follows:

1. The model with no tuition for students and thus wholly financed by the state. In this case, the state also gives aids to student. This model can explain the very high entry rates to tertiary education as bachelor's students (67%, well above the OECD rate, which is 60%). This is the case of European Nordic countries (see Chap. 6).
2. The model with high tuition fees, but accompanied by developing support to students. This model is typical of English-speaking countries (see Sect. 3.1.2) as well as of the Netherlands. The average entry rate to tertiary education for this group is even higher than that of the previous model (69%) and, thus, higher than the average level for OECD countries. As a consequence, the proportion of public subsidies in total expenditure on tertiary education is well higher than the OECD average (21%) in most countries of this group, the exception being the United States and Canada, where it is a little lower than this average.

   Even if students have to pay a high fee, however, foreign students in the United States are a proportion higher than the OECD average (70%, against 60%).
3. The model with high tuition fees, but a reduced support for students, i.e., a support lower than that for models 1 and 2. This model is typical of Japan and Korea (see Sect. 4.5). The students enrolled at tertiary level as bachelor's students in these two countries are 71% and 49% respectively. Public expenditure allocated to tertiary education in these two countries is very low.
4. The fourth group includes countries with low levels of tuition fees and less-developed student-support systems, where moderate tuition fees are charged in comparison with those in Models 2 and 3. It is typical of South European countries, as well as of Austria, Belgium, the Czech Republic, Ireland (see Chap. 4), Switzerland and Mexico.

1.2.2 *Methods and Models of Teaching and Learning*

As to *teaching methods*, Wikipedia, https://en.wikipedia.org/wiki/Teaching_method, says, 'the most basic teaching method is explanation', whose role is to make clearer the meaning of a method, or term or assignment). It can be used by both teachers and students to manifest comprehension. The explanation is important to make it clear both teaching and understanding. Students have an important role in explanation, because without student understanding, no explanation can be said to be clear.

There are a number of methods of teaching worth recalling: lecturing, demonstration, collaboration, debriefing, classroom action research.

The *lecturing* method is just one of several teaching methods, though it's usually considered the primary one. The lecturing method is convenient for the institution and cost-efficient, especially with larger classroom sizes. This is why lecturing is the standard for most college courses, when there can be several dozens of students in the classroom at once; lecturing lets professors address the most people at once, in the most general manner, while still conveying the information that they feel is most important, according to the lesson plan. While the lecturing method gives the instructor or teacher chances to expose students to unpublished or not readily available material, the students plays a passive role which may hinder learning. While this method facilitates large-class communication, the lecturer must make constant and conscious effort to become aware of student problems and engage the students to give verbal feedback. It can be used to arouse interest in a subject, provided that the instructor has effective speaking skills.

Demonstrating is the process of teaching through examples or experiments to prove a fact through a visual evidence and associated reasoning.

Collaboration allows students to actively participate in the learning process by talking with each other and listening to the opinions of the others. Classroom discussion is the most common type of such collaboration and can avoid intimidation.

Debriefing is a critical part of a study or experiment between human participants. It starts from the students' experiences and facilitates their reflection and feedback.

Classroom action research is a method for working in the classroom and for improving students' learning.

As to *teaching models*, we must first say that a teaching model is a set of guidelines for designing educational activities or a plan that can also be utilized to shape courses of studies, to design instructional material and to guide the principles of instruction to be followed by teachers for enabling learning by their students. These principles are determined either by the object of teaching or by the type of the learner.

They can be either authoritarian (teacher-centered) or authoritative (student-centered). In the first approach teachers pass knowledge and information to students, who are almost 'empty vessels' to be filled with. In this model, teaching is viewed as an entity separate from assessment. In the student-centered approach to learning, teachers are the authority figure, but both teachers and students play an active role in the learning process. The teacher measures student learning and overall comprehension of material through formal and informal assessments. Teaching and assessments are connected; student learning is continuously measured during the teacher's instruction.[4]

Models of teaching specify the kind of learning outcomes expected from the student, or his/her performance after the teacher's instructions. Therefore, these models select the instructional material and the appropriate instructions for students.

Teaching models differ also with respect to their inspiration. From this point of view there are: historical teaching models (Socratic, classical humanistic model

[4] One of the first contribution on this is by Passi et al. (1991).

and personal development model); philosophical teaching models (the impression model—such as that propounded by John Locke—the insight model and the rule model propounded by Kant); the psychological teaching models (with the basic teaching model, the interaction model, the school learning model and the computer based model); and, finally, the modern teaching model (with the group investigating model, propounded by John Dewey, the jurisprudential inquiry model, the social enquiry model, the role playing model and the laboratory method model).

From a more practical point of view, i.e., if we look mainly at the instruments used in teaching, we have:

1. Direct instructions. In this model the teacher gives information to the students directly, possibly with the help of some media, and answers questions raised by them. This is a model used in many high schools. The teacher will finally evaluate what students have learned.
2. Lecturing. The teacher gives lectures and provides information on facts and concepts face-to-face. He later checks what students have learned orally or by written tests. This is the more traditional teaching model, still followed in elementary schools.
3. Reference to online learning
4. Students form groups and cooperate together. They collaborate one with another to improve their knowledge and skills and teachers only provide some information.
5. Seminar and project-based learning. Students can work independently and learn through different tasks assigned to them, guided by teachers, in completing a product or doing other activities and designing a seminar.

References

Harmon, C. (2017). How effective is compulsory schooling as a policy instrument? *IZA World of Labor*.

Porcher, P (2020). State and school in Europe (Nineteenth–Twenty-first Century). In *Encyclopedie d'histoire numerique de l'Europe*. ISSN: 2677-6588. https://ehne.fr/en/node/14111

Radó, P. (2019). *Market reforms in the Hungarian school system: impact of changes in the ownership structure. NESET ad hoc question No. 2.*

OECD (2017). *PISA in Focus 2017/76*.

OECD (2022). *Education at a glance 2022: OECD indicators*. OECD Publishing, Paris. https://doi.org/10.1787/3197152b-en

OECD (2020), *PISA 2018 results (Volume V): Effective policies, successful schools, PISA*. OECD Publishing, Paris. https://doi.org/10.1787/ca768d40-en

Passi, B. K., Singh, L. C., Sansanwal, D. N. (1991). *Models of teaching: Report of the three–phased study of CAM and ITM*. Independent Study, National Council of Educational Research and Training, New Delhi.

Chapter 2
Short History of Education

Introduction

A historical analysis of education can be useful to know the path followed by the various peoples in the past periods, before the Common Era (Sect. 2.1) or this Era (Sect. 2.2) for giving an account of the roots of our present educational systems. Contemporary systems are indeed completely different from those existing in the past and, understandably, have registered a marked improvement on these, especially insofar as the diffusion of education is concerned.

2.1 Education Before the Common Era

In primitive civilisations education was the process of transmission of the culture of one generation to another in order to guide children to becoming good members of their tribe.

The earliest formal school was probably established in Egypt's Middle Kingdom, about 4000 years ago, for privileged male youth (one for scribes and the other for priest trainees), who were supervised by governmental officials and priests.[1] Pupils started at the age of 5 and learned reading and writing until the age of 16 or 17.

Almost at the same time, in Mesopotamia, a kind of education was developed quite similar to that of the Egypt for teaching the cuneiform script only to the sons of upper class of scribes and priests in order to perpetuate this highly developed civilisation. The objects of teaching were basic reading, writing, and religion at the beginning; later, they were extended to higher learning in law, medicine, and astrology. In any case girls were initially excluded, but later the syllabic script became more widespread and large part of the Mesopotamian population became literate. In

[1] The latter were a powerful intellectual elite in the Egyptian theocracy serving also to avoid political and cultural diversity, whereas the scribes served as professional copyists of manuscripts.

N. Acocella, *Understanding Education in the European Union*,
SpringerBriefs in Education, https://doi.org/10.1007/978-3-031-62065-2_2

Babylonian times there were libraries in most towns and temples. Rigid methods and severe discipline were applied in order to achieve uniformity in cultural transmission.

In ancient China and India too there were many schools.

In China education was ensured even before the invention of paper and the writing brush, in order to serve the needs of a simple agricultural society with the family as its basic social organisation. Oral instruction and teaching by example were the chief methods of education. Only the sons of nobles were initially educated, but later education became more widespread. In North China, more than 3000 years ago with the emergence of the Shang era complex educational practices were used. Education had a markedly secular and moral character having the purpose of developing a sense of moral sensitivity and duty toward people and the state. In North China, complex educational practices were in effect at a very early date. In fact, every important foundation of the formation of modern Chinese character was already established, to a great extent, more than 3000 years ago. Even in the early stage of civilisation, harmonious human relations, rituals, and music formed the curriculum. Formal colleges and schools probably antedate the Zhou dynasty of the 1st millennium BCE, at least in the imperial capitals. Local states probably had less-organised institutions, such as halls of study, village schools, and district schools. With regard to actual methods of education, ancient Chinese learned from bamboo books and obtained moral training and practice in rituals by word of mouth and example. Rigid rote learning, which typified later Chinese education, seems to have been rather condemned. Education was regarded as the process of individual development from within.

The purpose of education was of a social nature. In fact, it tended to develop moral sensitivity and loyalty toward people and the state, by teaching of harmonious human relations, rituals, and music.

More than 3000 years ago elementary and advanced schools were established in the main cities, initially for the sons of the nobility and later also for the common people to study rituals, music, archery, charioteering, writing, and mathematics. Girls studied chiefly homemaking and the feminine virtues ensuring the stability of the family system.

In India one of the most ancient civilisations in the world was established beginning almost 3000 years ago, due mainly to religious people, first the Brahmans, a group of intellectuals and priests, who were followed by the Kshatriyas and others. They impressed an important stimulus to education. At the beginning the children received elementary education at home and were then hosted in a home situated amid sylvan surroundings. In India there were some centres where Indian religions, Indian mathematics and Indian logic were taught. From the fourth to the eighth century the universities of Nalanda and Valabhi were founded and there was a rise of Indian sciences, mathematics, and astronomy, which attracted foreign students (Britannica, https://www.britannica.com/topic/education/Education-in-classical-cultures.)

In ancient Israel education was mainly familial, with the mother teaching the very young and the girls and the father responsible for moral, religious, and handcraft instruction of sons. The synagogue was also a school, where elementary and secondary or advanced levels of education were imparted to young boys. In primitive cultures education was the process of transmission of the culture of one generation to another in order to guide children to becoming good members of their tribe.

In Greece education was essential for the formation of a person's identity; it was both public and private and progressively more widespread, mainly for young males and non-slaves. In ancient Rome education was not diffuse, relying on *patria potestas* and most people were illiterate. Only later, in the Republican and the Empire time it became more widespread. The Roman education system was based on the Greek system and was not free.

Passing now to the New World, an important role was played by the Maya, Aztecs, and Incas. The Maya had a complex writing system, their knowledge of astronomy and mathematics was extraordinary and their calendar was so important as to surpass the accuracy of the Julian calendar. Similarly impressive were the sophistication of the Incas' calendar and their highway construction. In terms of construction, the magnificent temples of the Aztecs must be recalled.

In the Andean Inca empire too education had a relevant role. The Incas, similarly to the Aztecs, did not possess a written or recorded language and largely depended on oral transmission for maintaining and preserving their culture. Each individual's place in the society was given by strict rules. At five years of age children were taken over by the government, which supervised their socialisation and vocational training.

The Aztec Triple Alliance was the first state to implement universal compulsory education (around the year 1500). At the age of 15, all boys and girls went to school, whereas before that age education was responsibility of parents. However, girls were educated only in the crafts of home and child raising and not in reading or writing. It is important, however, to note that neither the Aztec nor the Inca used writing. The former had only pictographic and ideographic written records. The latter was a specific form of cuneiform tablets or painted symbols on papyrus.

2.2 Education in the Common Era

In Europe, most of the first Christian adherents were poor and illiterate, but education rose during the second century among leading citizens. These wanted their children to possibly have a better education, but there was a scarcity of schools, except for non-Christian culture. Thus, most Christians sent their children to the secular schools. In the process, also Christians set up catechetical schools, among which the most famous one was the one at Alexandria in Egypt.

Barbarian invasions entailed a breakup of the educational system developed by the Romans, even if Roman teaching and learning continued.

Later schools began to be established to some extent in the monasteries (as in the Monte Cassino monastery in Italy, which was founded in 529 by Benedict of Nursia) and in the cathedrals, which first specialised in the development of the clergy and later admitted young laypeople. In Ireland, first, and in England later there was a rising interest in education. The Carolingian renaissance has been represented as the renewal of Western culture, as the clergy was severely educated and the possibility arose of providing a more advanced training also for a few laymen. It was perhaps to meet this modest need that a school grew up within the precincts of the emperor's palace at Aachen. Apart from this, a few seats of culture were only here and there.

In the 'renaissance' of the twelfth century there was a rediscovery of the process of transformation of studies originating in the eleventh century. Contact with the East—deriving by trade and the Crusades—and the Moors in Spain stimulated intellectual life. 'Arabic renderings of some of the works of Aristotle, together with commentaries, were translated into Latin, exercising a profound influence on the trend of culture' (see Britannica, Education—Medieval, Monastic, Literacy).

All in all, during the Middle Ages, in Europe there were no public schools and the literacy rate was very low. Literacy was tied to the social status and thus reserved to the privileged elites, such as the nobility and wealthy landowners. The dominant role of the Church and specifically of the Catholic Church in Europe—but also in the Muslim world—and gender disparities largely influenced education, whose content was religious instruction, together with basic training in arithmetic, writing, and grammar. Thus, education perpetuated social hierarchies, with limited opportunities for the majority of people to have access to knowledge and learning.

However, in this period universities and other centers of higher education and research—such as the first universities of Bologna—the Alma Mother—founded in 1088, Paris—known metonymically as 'the Sorbonne', active from 1150—and Oxford—founded as early as 1096—were established.

Another key development was the creation of guilds, i.e., of associations of skilled craftsmen and merchants controlling the practice of their trades and responsible for vocational education of new members.

The protestant Reformation—under the impulse of Martin Luther—followed in Europe in around the same years, as the German Duchy of Württemberg established a compulsory education system, but only for boys, in 1559. The religious reforms introduced by Martin Luther, John Calvin, and others put emphasis on the importance of the state in the organisation of the educational system for children of all social levels, not simply of the richest. In the seventeenth and eighteenth century the system was introduced for both sexes by the German Duchy of Palatinate-Zweibrücken and by the city of Strasbourg. In 1763–5, Frederick the Great implemented a modern compulsory education system in Prussia to children from the age of 5 to that of 13–14. His minister founded a chair of pedagogy, aiming at improved education of teachers.

A few years later it was the turn of Austria, Hungary and the Czech lands for elementary education. The Prussian model gradually spread to other countries, being adopted by Denmark-Norway and Sweden, and Finland, Estonia and Latvia within the Russian Empire. Later, a system of education was introduced in England and Wales and France.

In the seventeenth century it was also the turn of Scotland, where a local land-based tax provided the required funding. As a consequence of the Scottish Reformation School, an Establishment Act of 1616 commanded every parish to establish a school for everyone, paid for by parishioners and supervised by bishops. There were then various decisions, partially reversing the initial one, but in 1696 a new Act re-established the compulsory provision of a school in every parish, with a system of implementation safeguard. Then Scotland was the first country in Europe with national compulsory education.

Education was very important for the leaders of the Counter-Reformation and, in order to educate children priests were needed. However, even if in theory education should have been universal, in practice it was almost entirely directed at the ruling classes. The most original contribution of the Counter-Reformation was the diffusion of ordinary verbal expressions rather than formal writings and reading.

During the eighteenth century, the education of girls was the specific object of the efforts of Madame de Maintenon, who became the wife of Louis XIV and founded a higher school mainly for orphan girls descending from noble families, where they were taught not only reading and writing, but also the basics for their future lives as wives and mothers.

Until the eighteenth century, only ecclesiastical schools and private persons schools existed in Russia where boys were taught reading, writing, arithmetic, singing, and religion. State-owned schools of mathematics, navigation, artillery, and engineering were established in Russia by Peter the Great in order to improve state administration and develop mining and industry. Then in 1786 Catherine II issued a statute for schools, according to which two-year courses were started in every town. In addition, a five-year course had to be established in every provincial town. Alexander I issued new statutes establishing state-controlled and free education, through parochial schools in rural areas, district schools of urban areas and provincial upper schools for the instruction of civil servants.

In the United States, following Luther and other Reformers, the Separatist Congregationalists obliged parents to teach their children how to read and write and the Massachusetts School Laws in the decade after 1640 moved toward compulsory education. In particular, hiring a teacher or establishing a school—according to the extension of local communities—were required. Secondary education became democratized in the United States in the early twentieth century, whereas in Europe it was attended by a minority of pupils until after the Second World War. In 1636, the Massachusetts General Court appropriated £400 for founding a college and two years later John Harvard, a Puritan minister, left some £800 to the school which therefore was given his name and accumulated 400 books of faith, which were left to Harvard University, initiating the library's collection. Later, in 1751 in Philadelphia Benjamin Franklin founded the Philadelphia Academy, which differentiated itself from other academies for the promotion of 'useful learning', including modern languages, history, geography, chronology, navigation, mathematics, natural and applied science.

As underlined by Porcher (2020), the spread of French revolutionary ideals, together with the desire to prevent new unrest and the need to train skilled labour, and construct nation states based on a shared culture made school to become a crucial issue in the nineteenth and twentieth centuries.

Even if inspired by the French revolution ideals, after Napoleon France was slow to introduce compulsory education, due to conflicts between the state and the Catholic Church. Finally, the Guizot law of 1833 provided education for boys. The Jules Ferry

Laws in 1881–82, extended primary education free for girls and boys until the age of 13 (raised to 14 in 1936 and to 16 in 1959).

Primary education became compulsory almost everywhere in Europe, in some cases basing on the Prussian model, in particular in: Austria in 1774, Greece (1834), Denmark and Iceland (1837), Sweden and Norway (1842), Spain (1857), Italy (1859), the canton of Geneva (Switzerland) and Scotland (1872), the United Kingdom (up to the ages of 11, 12 and 14, respectively in 1880, 1899, 1918), Denmark, with varying lengths in cities and countryside.

Compulsory schooling was often entirely free as in Sweden and Norway (1842), the canton of Geneva (1847), France (proclaimed in 1794, applied in 1882), and Greece (1895), or was partially free, as in Italy (1859), England (1891), and Denmark (1899). In Spain, free schooling was proclaimed in 1821, but was only partially implemented in 1907, when the state provided free schooling for poor pupils. In the United States education was a matter of single states. This explains why Massachusetts was the first state to pass a compulsory universal public education law in 1852 requiring every town to create and operate a grammar school, but the last state to enact a compulsory attendance law as late as in 1918 was Mississippi.

In the Soviet Union, a law providing compulsory education was implemented in 1930. In China nine-year compulsory education was formally established in 1986, which was very widely observed in a few years.

In Sweden and Norway, secondary schools were introduced, but there was no much attendance in the first years.

In the United Kingdom, secondary education was largely private. Public schools made up a small group of private and elitist secondary institutions that recruited nationwide, and were closely linked to Oxford and Cambridge Universities. The education of the British elites was highly homogenous, an expected result in a centralised system. The vast majority of secondary schools consisted of endowed secondary schools and those founded by local authorities, both of which were subsidized by the government and subject to inspection. There was no exam comparable to Prussia's *abitur* (1788), France's *baccalaureat* (1808), Denmark's *studentereksamen* (1850), Spain's *bachillerato* (1857), or Italy's *esame di maturità* (1923). In the early twentieth century, for every 1,000 primary ones there were 20 secondary pupils in Spain, 15 in France, and less than 5 in Italy. In Spain 78% of secondary schools were private, whereas 79% of schools for primary education were public.

Today, compulsory education is widespread in most countries. UNESCO has calculated that in the next 30 years the number of people who will receive formal education will be higher than in the whole human history thus far, also because of the growth of population. In addition, the percentage of illiteracy will further decrease, after a fall from 36% in 1960 to 25% in 2000, with the highest fall in less economically developed countries (LEDCs), which surpassed the fall of more economically developed countries.

Among developing countries, illiteracy and percentages of people without schooling stood in 2000 at about half the 1970 figures. Among developed countries, figures about illiteracy rates differed widely.

Since the mid-twentieth century, the various societies around the globe have undergone an accelerating pace of change in the economy and technology. The effects of changes on the workplace, and thus on the demands on the educational system preparing students for the workforce, have been significant. Beginning in the 1980s, government, educators, and major employers issued a series of reports identifying key skills and implementation strategies to steer students and workers towards meeting the demands of the changing and increasingly digital workplace and society. Twenty-first century skills are a series of higher-order skills, abilities, and learning dispositions that have been identified as being required for success in the twenty-first century society and workplaces by educators, business leaders, academics, and governmental agencies. Many of these skills are also associated with deeper learning, including analytic reasoning, complex problem solving, and teamwork, compared to traditional knowledge-based academic skills.

In 2021, instead, there were 18.9 million pupils in lower secondary education and 18.1 million in upper secondary education in the EU, but in most Member States of the European Union the latter outnumbered the former, the main exception being Germany and France. By contrast, in most of the enlargement countries (both candidate countries—Bosnia and Herzegovina; Montenegro; Moldova, North Macedonia; Albania; Serbia, Turkey and Ukraine—and potential candidates, i.e., Georgia and Kosovo), for which data are available there were more students in lower than upper secondary education (Eurostat, Secondary education statistics, https://ec.europa.eu/eurostat/statistics-explained/index.php?oldid=307350).

For EU Member States, the number of post-secondary non-tertiary pupils in 2021 compared with the number of upper-secondary pupils ranged from 30.3% in Germany to less than 2.0% in Portugal, France, Malta, Bulgaria and Italy. As noted above, this education level does not exist in five Member States.

In the EU, the vast majority (85.9%) of lower secondary education pupils in 2021 were taught in public institutions, as defined in the UNESCO, OECD, Eurostat (UOE) methodology. Across the vast majority of EU Member States at least two thirds of lower secondary school pupils were educated in public institutions, with the only exceptions of Malta (54.9%) and Belgium (39.7%).

More than 99.0% of all post-secondary non-tertiary students were educated in public institutions in Ireland, Luxembourg, Portugal and Lithuania; similar was the situation in countries candidate to the entry in the European Union such as North Macedonia and Serbia.

All post-secondary non-tertiary students—i.e., the higher technical education and training system managed by the State and the vocational training system managed by the Regions—were educated in the private sector in Italy, with particularly high shares also observed in Spain (95.3%) and Poland (89.3%). The percentages in Bulgaria and Belgium were around 75%.

Reference

Porcher, P. (2020). State and school in Europe (Nineteenth–Twenty-first Century). In *Encyclopedie d'histoire numerique de l'Europe*. ISSN 2677-6588. https://ehne.fr/en/node/14111

Chapter 3
Global School Systems

Introduction

This chapter offers a general view of the various ways schools are organised in various parts of the world: apart from Sect. 3.1, which gives a panoramic view of the various organizational systems, the chapter deals with schools: in developed English-speaking countries (Sect. 3.2); in Latin American and the Caribbean (Sect. 3.2.2); Africa (Sect. 3.2.3) and Asia (Sect. 3.2.4).

3.1 Schools Throughout the World

A first indication of the state of schooling in the various continents can be derived from the following Fig. 3.1.

Figure 3.1 shows that in most continents the number of children out of school has dramatically diminished in most continents and regions since mid-1980s, with the exception of Sub-Saharian Africa, where it has been oscillating, with a dramatic rise in the first 10 years, followed by a sensible decline until 2005–2007 and a new rise afterwards. Obviously, similar data give only a very rough indication of the situation, as they offer information about values in absolute terms.

As Roser, Ortiz-Ospina (2013)[1] say, more than half (almost 60%) of primary-aged children not in school were in Sub-Saharan Africa in 2019 and South Asia accounted for another 19%. Percentages at a given time, without the indication of their time trend, can be misleading, however, since in Sub-Saharan Africa the total number of primary school children out-of-school had fallen by 10 million since the late 1990s to the end of the first decade of the 2000s, but have risen more recently, as said.

An overview of the situation as to the average years of schooling in the various continents in 2020 and the change after 1870 is offered by Table 3.1.

[1] See also the updated versions of their paper in the internet entry on World Population Growth.

N. Acocella, *Understanding Education in the European Union*,
SpringerBriefs in Education, https://doi.org/10.1007/978-3-031-62065-2_3

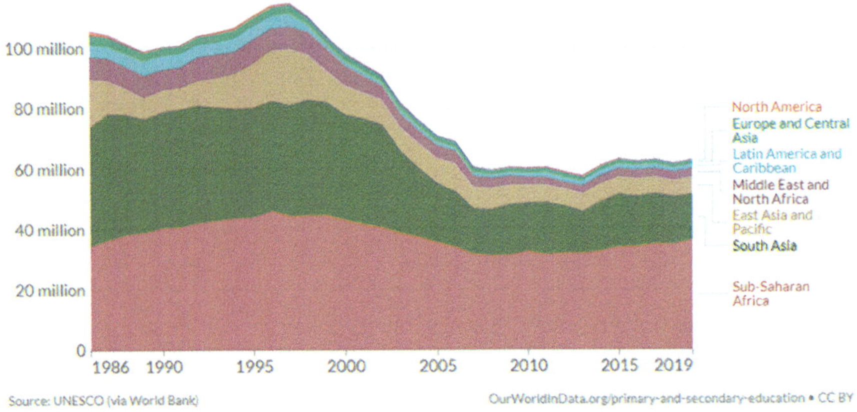

Fig. 3.1 Primary-school-age children who are out of school by world region, 1986–2019. *Source* Our World in Data, https://ourworldindata.org/grapher/out-of-school-children-of-primary-school-age-by-world-region

Table 3.1 Average years of schooling in the various continents in 1870 and 2020. *Source* Wikipedia, Education in Africa, https://en.wikipedia.org/wiki/EducationinAfrica

Africa	0.1	6.7	+ 6.6	+ 11.363%
Asia	0.0	8.6	+ 8.5	+ 21.246%
Europe	1.2	11.9	+ 10.7	+ 897%
North America	3.1	11.9	+ 8.7	+ 277%
Oceania	1.9	10.7	+ 8.9	+ 476%
South America	0.2	8.9	+ 8.7	+ 4.834%
World	0.5	8.8	+ 8.4	+ 1.805%

Source Unesco, OurWorldInData.org/global-education[2]

Unesco's World Inequality Database on Education can offer a complement to these figures. It gives a number of indicators as to school attendance and learning, ranging from completion rates in the various levels of education (from the primary to the upper secondary levels) and the number of out-of-school children to learning achievements in the various fields.[3]

To indicate the status of the various countries in terms of education, we will use the *Education Index*, which is a component of the Human Development Index. It is measured by combining average years of schooling for adults with expected years of schooling for children, with a 50% weight for each. Figure 3.2 shows the Index for all countries in the world.

[2] The various columns indicate: Continents, Data for 1870, Data for 2020, Absolute change, Relative change.

[3] It is interesting that visiting the UNESCO site at https://ourworldindata.org/grapher/mean-years-of-schooling-long-run?time=2017 one can have a view of the evolution of schooling since 1870.

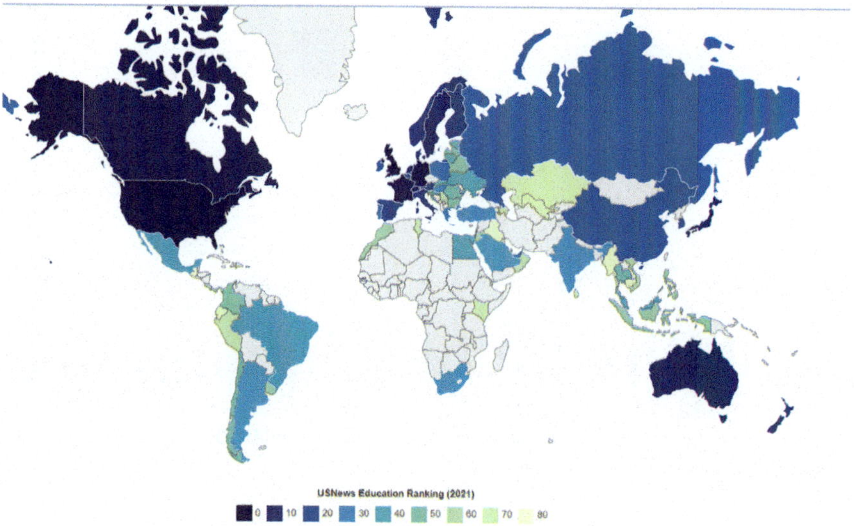

Fig. 3.2 Education Index for all countries in the world, 2021. *Source* UNESCO, World Population Review, https://worldpopulationreview.com/country-rankings/education-rankings-by-country. *Note* The Index is higher the darker the colour depicting the country in the figure

For example, here we indicate the value of the index for the first 10 countries in the world: Germany, 0.940; Australia, 0.929; Denmark, 0.920; Ireland, 0.918; New Zealand, 0.917; Norway, 0.915; United Kingdom, 0.914; Iceland, 0.912; Netherlands, 0.906; Finland, 0.905. By contrast, the last 5 countries had the following index: Central African Republic (56%), Ethiopia (39%), Mali (31.1%), Burkina Faso (28.7%), Niger (28.7%).

With specific reference to the trend in the share of tertiary educated young people since the beginning of the century, Fig. 3.3 shows its marked increase in OECD countries.

Despite the fact that literacy is today higher than ever and younger generations are certainly more educated, as data on the literacy rates by age groups show, we still face now important challenges in many developing countries. In these countries the literacy rate for the overall population will continue to increase. Therefore, estimates and projections across countries of the share of individuals who have no education show that only a few countries (Burkina Faso, Ethiopia, Guinea, Mali and Niger) will have a rate of no education above 20% in 2050, as said by The International Institute for Applied Systems Analysis (IIASA) (Roser & Ortiz-Ospina, 2013).

Another aspect of international comparisons refers to gender parity. According to UNESCO, Latin America and the Caribbean are second in the world for gender parity in higher education, after Oceania, as can be seen from Fig. 3.4.

Note: Starting from the top in 2020, curves indicate the following regions: Oceania, Latin America and the Caribbean, Europe and North-America, Eastern and

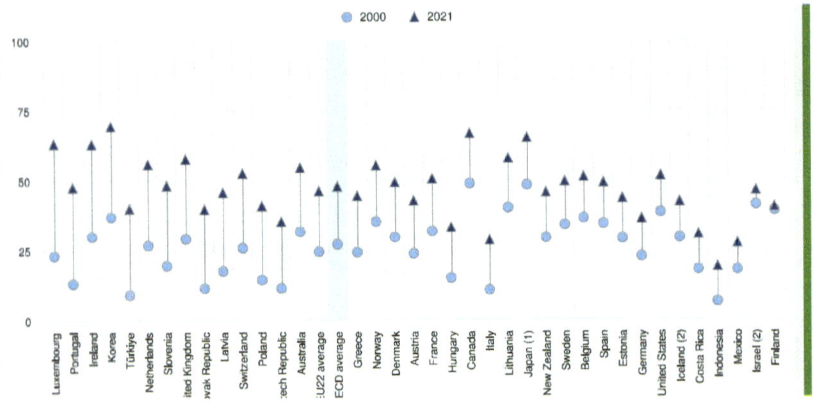

Fig. 3.3 Trends in the share of tertiary-educated 25–34 years-old (2000 and 2021), %, OECD countries. *Source* OECD (2022)

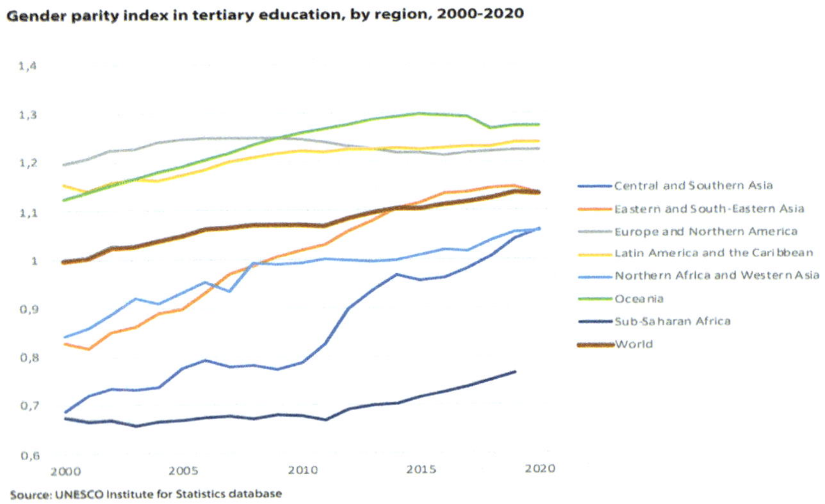

Fig. 3.4 Gender parity index in tertiary education, by region, 2000–2020. *Source* UNESCO

South-Eastern Asia, World, Northern Africa and Western Asia, Central and Southern Asia, Sub-Saharan Africa.

For a longer span of time, Africa, South America, and Asia have been lagging behind the other continents, in terms of the distance between females and males as to the average number of years for which people aged 15–64 have participated in formal education since 1870 or will do so in the next future. The ratios are now very close to 100% and females will practically reach males in 20 years time, as shown by Fig. 3.5.

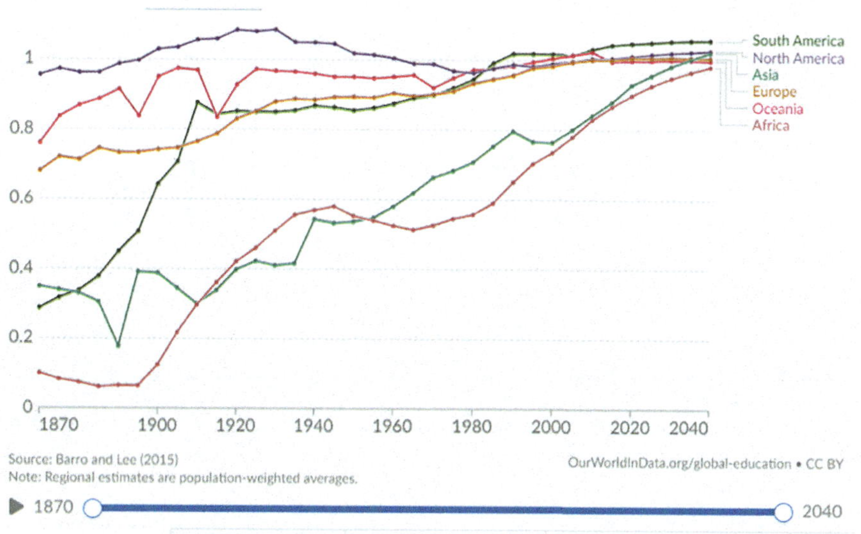

Fig. 3.5 Gender ratio for average years of schooling, 1870–2040. *Source* Our world in data, https://ourworldindata.org/grapher/gender-ratios-for-mean-years-of-schooling. *Note* Starting from the top in 1870, regions are: North America, Oceania, Europe, Asia, South America and Africa

The preceding visualisation shows that, despite the fact that literacy is today higher than ever, there are still important challenges in many developing countries. However, data on literacy rates by age groups show that in most countries, and certainly in virtually all developing countries, there are large generational gaps: younger generations are progressively better educated than older generations. This indicates that in these countries the literacy rate for the overall population will continue to increase.

Figures from IIASA suggest that we should see rates of education increasing as the world develops.

While these projections entail a number of assumptions, the conclusion seems to be that by 2050 we can hope that most of the cross-country gaps in literacy will be closed.

As to a more limited age class, that of 25–34 years-old, a gender gap in educational attainment is opening. In fact, on average, women are now 57% of younger adults with at least a bachelor's or equivalent degree in all OECD country except Japan.

Female-to-male ratio of the average number of years people aged 15–64 participated in formal education. Values below 1 mean that fewer women were educated.

Education has a positive economic impact, since additional earnings can derive from it, as indicated in the following Fig. 3.6 with reference to various world regions.

Looking at the relationship between the GDP per capita and average learning outcomes is of some interest. In fact, Fig. 3.7 shows that the former is higher when the latter is high too. Among the countries with the highest per capita GDP there are Singapore, Ireland, Luxembourg, the United States and Japan, which are also

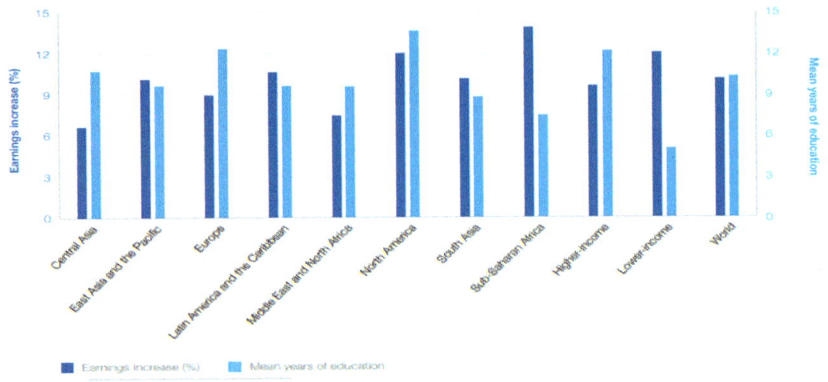

Fig. 3.6 Earnings increase (%) and mean years of education, various regions. *Source* World Bank (2017)

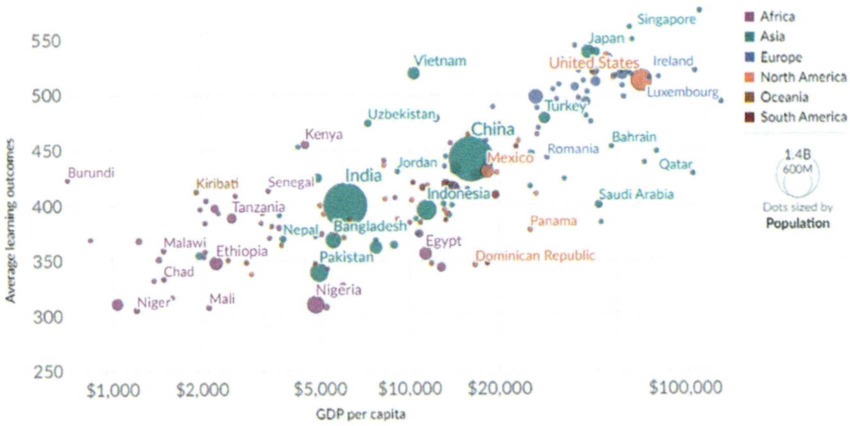

Fig. 3.7 Average learning outcomes and GDP per capita in the various countries, 2020. *Source* Our World in Data, https://ourworldindata.org/global-education

the countries with the highest average learning outcomes, with the exception of the United States, which was only 31st in the PISA 2022 tests.

3.2 Regional Overview

3.2.1 Schools in the United Kingdom, North America, Australia and New Zealand

- Schools in the United Kingdom

The United Kingdom has a strong state-funded school system that provides free education, from preschool to higher education, for all people having the right of abode. In 2021, 10.5 million students attended state schools across the United Kingdom, according to national statistics. About 93% of children between the ages of 3 and 18 were in education in state-funded schools without charge (other than for activities such as swimming, cultural visits, theatre visits and field trips for which a voluntary payment could be requested) and limited charges at state-funded boarding schools. Parents can request financial support to cover uniform costs, physical education (PE) kits, and transport to school, but each region of the United Kingdom has its own scheme.

There are also many independent schools across the United Kingdom. These include 'private' (which are called 'public' schools there), international, and boarding schools. Among them Eton, Harrow, and Winchester are some of the United Kingdom's best-known 'public' schools. The government does not fund, nor partially subsidise these private schools. Parents therefore have to pay tuition fees, which can be expensive. Across the United Kingdom, there are more than 2,500 independent or private schools (including international and boarding schools). Each year they teach around 615,000 students from the age of 4 to 18 for primary and secondary education.

Contrary to what happens in many European counties, homeschooling (or home education or elective home education, EHE) is a part of the United Kingdom education system. In fact, all four education departments offer support and resources to ensure high-quality tuition for home-educated children.

Parents can homeschool their children full-time or part-time, even if they do not need to follow the national curriculum. However, the local councils regularly review their progress. If it is found lacking, they may order parents school attendance, which means that the child has to attend the local school.

Full-time education is compulsory for all children aged 5 to 18 (but schooling is compulsory only up to the age of 16) and ends with an exam called the GCSE (General Certificate of Secondary Education, i.e., a certificate introduced in the UK in September 1986 to establish a national qualification for pupils leaving school at 16). In the United Kingdom the compulsory education system goes through four stages: preschool, primary school, secondary school, and high school. At the end of secondary school students are required to pass written exams in 9 subjects (among which Mathematics, English, and Science are core compulsory subjects) to attain their GCSE. This is in Year 11 in England and Wales and Year 12 in Northern Ireland. In Scotland the National 5 exams are the equivalent of the exams in other regions. In 2021, a grade 5 (strong pass) or more was achieved by 51.9% of pupils across the United Kingdom in their GCSE exam for Mathematics.

Children begin their primary school at the age of 5. Pupils are not given specific grades, as parents receive only a description showing the strengths and weaknesses of their sons and daughters, and their learning progress. Students move on to the next level of education, regardless of the results of their exams.

Higher education often begins with a three-year bachelor's degree, followed by postgraduate degrees, which include master's degrees, and doctoral level research degrees lasting at least three years.

British universities have been leaders in academia for centuries and two of them (Oxford and Cambridge) are among the world's top three; in addition, a further 26 are ranked among the global top 200 by the Times Higher Education, World University Rankings 2023.

Generally, the quality of the United Kingdom education system is good. For instance, the United Kingdom performed above the OECD average in the 2022 PISA tests for reading, mathematics, and science.

This is in Year 11 in England and Wales and Year 12 in Northern Ireland. Mathematics, English, and Science are core subjects, which every student must take. The equivalent in Scotland is the National 5 exams. In 2021, 51.9% of pupils across the United Kingdom achieved a grade 5 (strong pass) or above in their GCSE exam for mathematics.

Education in the United Kingdom beyond these years is called 'further education' and typically runs until the student is 18 years old. At the end of this education, students typically complete the Advanced (A-Level) exams at the end of Year 13 (England and Wales) or Year 14 (Northern Ireland) to graduate. In Scotland, *the Highers* and *Advanced Highers* are optional secondary qualifications. Average A-Level results increased from C+ to B in 2021.[4] However, one must account for the impact COVID-19 had on schooling and assessments.

Across the United Kingdom, attending preschool or nursery is voluntary. Still, children above the age of three can access some free childcare hours per week.

Usually, the primary stage in the United Kingdom education system starts from four or five and covers about seven years of schooling. After primary school, students attend secondary school. Even though—as said—compulsory education includes about 12 years until the age of 16, most students complete 2 more years to apply to universities (Laas, 2023).

- Schools in the United States

In the United States, education is compulsory starting between the ages of five and eight and ending between the ages of sixteen and eighteen, depending on the state. There are elementary schools, middle or junior high schools, and high schools. Education is both public and private. Also individuals can provide education through homeschooling. Funding is provided mainly by the states and local governments (most of the $1.3 trillion expenditures in 2021), and to some extent by the federal government (only $260 billion, in the same year). Overall, spending by public governments was almost 93% in 2020, to be compared to that by private funds, equal to 7%. Comparable data for homeschooling are not available, for the rise of this form of education due to the pandemic. Expenditures per student are higher in the United States than for any other country. In 2014, the country spent 6.2% of its GDP on all levels of

[4] Consider, however, that in 2014 England changed its grading system – i.e. GCSE - from the A to G system to a 1 to 9 numerical scale, 9 being the highest.

education, a percentage higher than that of the OECD average of 5.2%. Expenditures per-pupil for primary and secondary spending in the United States was 34% higher than the OECD average (ranking 5th of 36 countries reporting data) in 2018; post-secondary 'per-pupil spending was double the OECD average (ranking 2nd), and the U.S. education sector consumed 6% of the U.S. GDP (ranking 6th). In the last decades, girls have generally outperformed boys on measures of grades across all subjects and graduation rates, which is a turnaround from the early twentieth century when the contrary happened. However, boys score higher on standardised tests than girls and are better represented in the more prestigious, high-paying STEM (i.e., jobs focusing on the academic disciplines of Science, Technology, Engineering and Mathematics) (see Wikipedia, Education in the United States, https://en.wikipedia.org/wiki/Education_in_the_United_States).

Educational standards for public schools are set by state governments, whereas private schools are free to determine their own curriculum and staffing policies.

The large majority (19 of the top 25) of the world's top universities are American and the U.S. has by far the most Nobel Prize winners in history, having 406 out of 989 total awards between 1901 and 2022. However, the average overall knowledge and skills of American student was ranked only 25th in the world by PISA. In 2018 the average American student scored 478, compared with the OECD average of 489 in Mathematics, overcoming the average in reading (505 against 487) and Science (502 against 489). In 2014, the U.S. Education has been rated as 14th best in the world by the Economist Intelligence Unit. PISA has ranked the knowledge and skills of American 15-years-old as 19th in the world in literacy, mathematics, and science with the average American student scoring 495, compared with the OECD average of 488. The average difference between advantaged and disadvantaged students in reading was 99 points, compared to an average of 89 in OECD countries. However, 10% of disadvantaged students were academically resilient (OECD average: 11%).

- Schools in Canada

In Canada education is for the most part provided publicly, being funded and overseen by provincial, territorial and local governments. After the kindergarten at ages 4 or 5, children attend the primary education, which is followed by secondary and post-secondary education and is compulsory up to the age of 16 or 18 according to the province.

In 2017, 68% of Canadians aged 25 to 64 attained good post-secondary education, against the OECD average of 44%.

The PISA Program ranks the overall knowledge and skills of Canadian 15-years-old as the sixth best in the world (better than those of the United Kingdom and the United States). Canadian students perform better than the OECD average in literacy, mathematics, and science (520, as compared with the OECD average of 487 in 2018). To reach these results, this country spent 6.0% of its GDP in 2016 for all levels of education—roughly 1.0 percentage points above the OECD average of 5.0% and a bit lower than the United States percentage. In 2017, 68% of Canadians aged 25–64 attained post-secondary education, well above the OECD average of 44% (Wikipedia, Education in Canada, https://en.wikipedia.org/wiki/Education_in_Canada).

- Schools in Australia

In Australia education is compulsory for 11 years between the ages of 4 to 6 and 15 to 17, depending on the state or territory and the date of birth, in government or non-government schools.

According to PISA 2018, Australia was ranked 21st, after Japan, the United Kingdom and the United States. Instead, the United Nations Education Index indicated Australia as the second-highest provider of education in the world in 2017. Regulation and funding of education are primarily the responsibility of the States and territories. However, the Australian Government also plays a funding role.

Approximately 60% of Australian students for primary and secondary education attended schools financed by the government. At the tertiary level too, the large majority of Australia's universities are public, and student fees are subsidised by a loan program whose repayment will be due when debtors reach a certain income level.

In 1966 the Australian Government signed the Convention against Discrimination in Education, which aimed to combat discrimination and racial segregation in the field of education.

In 2012 Australia was the third-largest provider of education to international students, after the United States and the United Kingdom, having also the highest ratio of international students per head of population in the world (Wikipedia, Education in Australia, https://en.wikipedia.org/wiki/Education_in_Australia).

- Schools in New Zealand

The education system in New Zealand implements the usual three-tier model including primary, secondary schools (from 5 to 9 years of age) and tertiary education. PISA 2018 ranked New Zealand 14th in the world, before the US. In 2018 the PISA program ranked New Zealand 12th-best at science, 12th-best at reading, and 27th-best in mathematics, but New Zealand's mean scores have been steadily dropping in all three categories. Most schools in New Zealand are owned and funded by the state, where teaching is according a national curriculum. Private schools are an exception, and they develop their own learning programmes. Private schools are mostly funded through charging parents with school fees and get limited government funding. 'The Human Rights Measurement Initiative found that as of 2022 New Zealand achieved 95.9% of what should be possible at its level of income for the right to education.' (Wikipedia, Education in New Zealand, https://en.wikipedia.org/wiki/Education_in_New_Zealand).

3.2.2 *Schools in Latin America and the Caribbean*

In order to understand the performance of schools in this region, one must first realise that now poverty has increased in Latin America with respect to twenty years ago, despite the region's relatively high economic growth. It is well known that illiteracy is frequent among the poor, which implies that generally poor's children have limited opportunities to learn and thus escape the poverty trap.[5] In fact, the poor have lower levels of education. A positive aspect of the situation is that almost 100% of children (but not all of them, even if universal free primary education is mandatory) are enrolled now in first grade, more, with respect to 60 years ago. In fact, only 60% of children were enrolled in first grade in the early 1960s, against 95% today.

In Latin America, the sharp inequalities in the distribution of income reflect themselves in equally sharp inequalities in the distribution of access to knowledge and skills.

Despite significant progress, education remains a challenge in Latin America. Children complete on average two more years of schooling than their parents' generation. Various types of administrative and institutional reforms have been implemented for enabling to reach higher education levels for communities having no access to education services in the early 1990s. However, 23 million children in the region between the ages of 4 and 17 are still outside the formal education system.

Considerable progress was made in expanding access to education in the past decades. Enrollment rates since 1960 to the end of the Nineties increased from 60 to 88% at the primary level, from 36 to 72% at the secondary level and from 6 to 27% at the tertiary level. Some countries (such as Mexico and Chile) have achieved a significant increase in the access to education (87% of the population today completing their primary schooling compared to 46.6% in 1980 in Mexico and almost 80% in Chile).

Often, the poor's children do not complete the schooling because their parents find it difficult to continue supporting their studies. The performance of these poor in tertiary education in Colombia and Mexico was found to be comparable to that of the rich according to Reimers (1999), but PISA 2015 found that in general poor students in the region still performed far worse than their wealthier peers (OECD, 2016).

According to UNESCO, CEPAL, UNICEF (2022), the goals established in 'Sustainable Development Goal 4 (SDG4) of the 2030 Agenda for Sustainable Development' were not met before the pandemic, and are even less so afterwards. Certainly, there has been an increase, on the one hand, in the adult literacy levels (and in general of the educational level achieved by the population together with an improvement in the access to pre-primary education, especially in rural areas and in the lowest income quintile of the population) and, on the other, of completion of secondary education, especially among the most vulnerable groups.

However, there has been a slowdown in the improvement of certain indicators—such as secondary education completion rates—and a stagnation in key indicators of access to primary and secondary education and in the assessments of the quality of learning, together with an increase in certain specific gaps at the tertiary level.

---

[5] A test of the close relation between education and income is offered by Brazil, where the poorest 40% of teenagers attend schools only for 4 years, while those in the top 20% of income distribution have about 80% of teenagers attending.

Many schools lack basic inputs, such as libraries, textbooks, science labs, meeting spaces for teachers, computer rooms, and gymnasiums. Similarly, a number of them have no access to potable water and electricity, or lack a drainage system and phone lines, or have an inadequate number of bathrooms (Wikipedia, Education in Latin America, https://en.wikipedia.org/wiki/Education_in_Latin_America).

With regard to financing for education, a key dimension of education policy and the trends identified before, 15 countries in Latin America and the Caribbean have reduced their public investment in education since 2015. This reduction intensified with the COVID-19 crisis, which in 2020 alone brought an economic contraction of 7.7% of GDP. Between 2015 and 2019, spending for education as a percentage of total public spending fell from 16.1 to 15.4% in the region. Relative to GDP, this percentage fell from 4.5 to 4.3%.

Given all these facts, it comes as a consequence that almost two-thirds of Latin American students do not achieve satisfactory reading and mathematics scores and there is a significant learning gap between students from different socio-economic backgrounds.

As to the gender of young people attending school, more than 60% women go to college today compared to less than half of men and there has been a substantial increase since 1970 when only 5% of women were educated beyond secondary school (see Fig. 3.8).

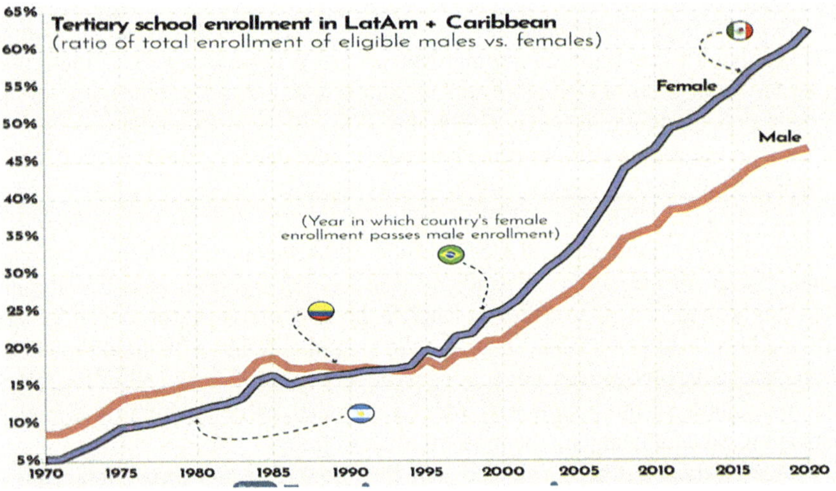

Fig. 3.8 Tertiary school enrollment in Latin America and the Caribbian by gender, 1970–2020. *Source* UNESCO, https://data.worldbank.org/indicator/SE.TER.ENRR.FE?locations=ZJ

3.2.3 Schools in Africa

In pre-colonial Africa there was no formal education, with very few exceptions – such as that of the ancient Egypt and some Sahelian kingdoms in West Africa. One of the first forms of higher education in Africa was the School of Holy Scriptures.[6] Apart from that, education was practically limited to oral transmission of notions by older members of the family.

Instability resulting from, and exacerbated by, armed conflicts in many regions of Africa as well as from the fallout of colonialism had an important influence on education in Africa. The Belgian king Leopold II prohibited access to higher education in his few colonies in this continent (the Republic of Zaire, now Democratic Republic of the Congo, and Ruanda-Urundi, now the Republic of Rwanda), whilst other colonial powers put in various barriers to access, such as the language of instruction—which had to be that of the colonizer—and limits on teaching curriculums, and ensuring the curriculum did not reflect any Afro-ethnicity. Specific cases of positive effect of colonialism can be found within these limits, whose respect could have been useful for better governing colonies.

Other negative influences derived from armed conflicts, famines, lack of drinking water, diseases and inequalities, being thus able to dictate what the people learned.

A decisive impact derived also from the conditions of weather, which caused frequent migrations, also of a temporary nature, that were of obstacle to education. Education organised by the state existed only in Sahelian states, due to their relatively higher wealth (Wikipedia, Education in Africa, https://en.wikipedia.org/wiki/Education_in_Africa).

Now of all regions, Sub-Saharan Africa has the highest rates of education exclusion. Over one-fifth of children between the ages of about 6 and 11 are out of school, followed by one-third of youth between the ages of about 12 and 14. According to the UNESCO Institute for Statistics (UIS)[7] data, almost 60% of youth between the ages of about 15 and 17 are not in school.

Without urgent action, the situation will likely get worse as the region faces a rising demand for education due to a still-growing school-age population. Education in Africa is a major priority for UNESCO.

Girls' education is one of the most important goals, since 9 million girls aged 6 to 11 (i.e., 23% of all) do not attend any school at all. The same situation is for 6 million boys (19%) (see UNESCO UIS, Education in Africa, http://uis.unesco.org/en/topic/education-africa).

---

[6] Some of the first forms of higher education in Africa were the School of Holy Scriptures in Ethiopia and Al-Azhar in Egypt. These became cultural and academic centers for many people who travelled there from all over the globe for knowledge and instruction.

[7] This institute develops indicators to better address education challenges. For example, it tracks the extent to which schools lack – in addition to teachers - basic 'amenities', such as access to electricity and potable water or good classroom conditions, availability of textbooks and existence of multi-grade classrooms, with seven out of ten countries facing an acute shortage of teachers. The Institute also produces a range of data on teachers' training, recruitment and working conditions.

Kaledzi (2022) confirms that nearly 60% of African youth aged 15–17 are not in school.

As to total population aged 15 or more, we have the following data of percentages of educated people (see Table 3.2). The following five countries are ranked highest: Seychelles, with 95.3% of educated people, Equatorial Guinea (95.3%), South Africa (95%), Mauritius (91.3%), Lybia (91%). The last five countries are instead: Burkina Faso (39.3%), Central African Republic (37.4%), Mali (35.5%), Niger (35.1%), South Sudan (34.5%).

In 2015, a set of Sustainable Development Goals was adopted by the United Nations for the year 2030. Among them, the fourth goal addressed education, aiming at ensuring 'inclusive and equitable quality education and promote lifelong learning opportunities for all' (United Nations, 2015, https://sdgs.un.org/goals).

For the large majority of countries, the percentages of education for men are much higher than those for women. Botswana, Eswatini (formerly Swaziland), Lesotho and Seychelles are the only countries where education is higher for females than for males. In a few other countries, such as Gabon, Namibia and South Africa, the percentages of the two sexes are very close.

3.2.4 *Education in Asia*

The Gross Enrollment Ratio (GER) is a component of the Education Index. It expresses the number of students enrolled in a given level of education as a percentage of the number of people within the official age for that level of education. The ratio can exceed 100% because some enrolled students may fall outside the official age range. Thus, differences between the two ratios show the incidence of over-age and under-age enrollments.

Table 3.3 below shows the GER for each level of education and each country in South Asia, East Asia, Southeast Asia, West Asia and Central Asia.

The last year for which data are available is shown in parentheses following each number in the table. If the year is the same as for the column to the left, the year is omitted.

Practically all Asian countries had 100% attendance for primary schools, the exceptions being Pakistan, Sri Lanka, Brunei, Azerbaijan, Jordan, Palestine, Syria, Turkmenistan, Uzbekistan. As for tertiary education, South Korea had the highest attendance (97%) and was followed by Turkey (79%), Mongolia (62%), Japan (61%) and Iran (58%).

As Asian nations compete in the global economy and aspire to join the developed nations, there is concern that their rates of education may not be keeping pace. Consider that Japan and China – which are certainly two of the most important economic players in the world - had rather low levels of tertiary education (respectively, 30% and 61%). By comparison, Gross Enrollment Rates for North America and Western Europe in 2013 were 84.3% for pre-primary, 101.1% for primary, 105.1% for secondary, and 76.6% for tertiary education.

Table 3.2 Percentage of educated people, male and female, in African countries, various years

Countries	Total population (%)	Male (%)	Female (%)	Year
 Algeria	81.4	87.4	75.3	2018
 Angola	71.1	82	60.7	2015
 Benin	42.4	54	31.1	2018
 Botswana	88.5	88	88.9	2015
 Burkina Faso	39.3	49.2	31	2018
 Burundi	68.4	76.3	61.2	2017
 Cabo Verde	86.8	91.7	82	2015
 Cameroon	77.1	82.6	71.6	2018
 Central African Republic	37.4	49.5	25.8	2018
 Chad	22.3	31.3	14	2016
 Comoros	58.8	64.6	53	2018
 Congo, Democratic Republic of the	77	88.5	66.5	2016

(continued)

Table 3.2 (continued)

Countries	Total population (%)	Male (%)	Female (%)	Year
 Congo, Republic of the	80.3	86.1	74.6	2018
 Cote d'Ivoire	89.9	93.1	86.7	2019
 Egypt	71.2	76.5	65.5	2017
 Equatorial Guinea	95.3	97.4	93	2015
 E ritrea	76.6	84.4	68.9	2018
 Eswalini	88.4	88.3	88.5	2018
 Ethiopia	51.8	57.2	44.4	2017
 Gabon	84.7	85.9	83.4	2018
 Gambia, The	50.8	61.8	41.6	2015
 Ghana	79	83.5	74.5	2018
 Guinea	39.6	54.4	27.7	2018
 Guinea-Bissau	59.9	71.8	48.3	2015
 Kenya	81.5	85	78.2	2018

(continued)

Table 3.2 (continued)

Countries	Total population (%)	Male (%)	Female (%)	Year
Lesotho	79.4	70.1	88.3	2015
Liberia	48.3	62.7	34.1	2017
Libya	91	96.7	85.6	2015
Madagascar	76.7	78.4	75.1	2018
Malawi	62.1	69.8	55.2	2015
Mali	35.5	46.2	25.7	2018
Mauritania	53.5	63.7	43.4	2017
Mauritius	91.3	93.4	89.4	2018
Morocco	73.8	83.3	64.6	2018
Mozambique	60.7	72.6	50.3	2017
Namibia	91.5	91.6	91.4	2018
Niger	35.1	43.6	26.7	2018

(continued)

Table 3.2 (continued)

Countries	Total population (%)	Male (%)	Female (%)	Year
 Nigei https://en.wileip edia.ora/wiki/Niger	62	71.3	52.7	2018
 Rwanda	73.2	77.6	69.4	2018
 Sao Tome and Principe	92.8	96.2	89.5	2018
 Senegal	51.9	64.8	39.8	2017
 Seychelles	95.9	95.4	96.4	2018
 Sierra Leone	43.2	51.6	39.8	2018
 South Africa	95	95.5	94.5	2019
 South Sudan	34.5	40.3	28.9	2018
 Sudan	60.7	65.4	56.1	2018
 Tanzania	77.9	83.2	73.1	2015
 Togo	66.5	80	55.1	2019
 Tunisia	81.8	89.6	74.2	2015

(continued)

Table 3.2 (continued)

Countries	Total population (%)	Male (%)	Female (%)	Year
Uganda	76.5	82.7	70.8	2018
Zambia	86.7	90.6	83.1	2018
Zimbabwe	86.5	88.5	84.6	2015

Source Wikipedia, Education in Africa, https://en.wikipedia.org/wiki/Education_in_Africa

Table 3.3 Gross Enrollment Ratio in the different levels of education, 2013, *Source* Wikipedia, Education in Asia, https://en.wikipedia.org/wiki/Education_in_Asia

Gross Enrollment Ratio: South Asia[1] Country	Population (millions) 2013	Pre-primary	Primary	Secondary	Tertiary
Afghanistan	30.6	1% ('03)	106% ('13)	54%	4% ('11)
Bangladesh	156.6	33% ('13)	114 ('11)	54% ('12)	13%
Bhutan	0.8	14% ('13)	107%	78%	11%
India	1,252.1	58% ('11)	114% ('12)	71%	25% ('13)
Maldives	0.3	83% ('07)	105%	72% ('04)	13% ('08)
Nepal	27.8	87% ('14)	133%	67%	17% ('13)
Pakistan	182.1	82% ('13)	92%	38%	10%

(continued)

Table 3.3 (continued)

Gross Enrollment Ratio: South Asia[1] Country	Population (millions) 2013	Pre-primary	Primary	Secondary	Tertiary
Sri Lanka	21.1	90% ('13)	98%	99%	19%
China	1,385.6	74% ('13)	126%	92%	30%
Japan	127.1	88% ('12)	102%	102%	61%
Mongolia	2.8	86% ('12)	109% ('13)	92% ('10)	62% ('13)
North Korea	24.9	NA	NA	NA	NA
South Korea	49.3	93% ('14)	100%	99%	97%
Taiwan	NA	NA	NA	NA	NA
Brunei	0.4	64% ('13)	94%	106%	25%
Cambodia	15.1	15% ('13)	125%	45% ('08)	16% ('11)
East Timor	1.1	10% ('05)	NA	NA	NA

(continued)

Table 3.3 (continued)

Gross Enrollment Ratio: South Asia[1] Country	Population (millions) 2013	Pre-primary	Primary	Secondary	Tertiary
Indonesia	249.9	51% ('13)	109% ('12)	83% ('13)	32% ('12)
Laos	6.7	26% ('13)	NA	NA	NA
Malaysia	29.7	84% ('12)	101% ('05)	71% ('12)	37%
Myanmar	53.3	9% ('10)	114%	50%	13% ('12)
Philippines	98.4	52% ('09)	107% ('13)	85%	34%
Singapore	NA	NA	NA	NA	NA
Thailand	67.0	119% ('13)	100%	86%	51%
Vietnam	NA	NA	NA	NA	NA
Armenia	3.0	46% ('13)	102% ('09)	97%	46% ('13)
Azerbaijan	NA	25% ('12)	98%	100%	20%
Bahrain	1.3	53% ('13)	104% ('99)	101% ('11)	40% ('14)
Cyprus	0.9 ('12)	78% ('12)	100%	95%	46%

(continued)

Table 3.3 (continued)

Gross Enrollment Ratio: South Asia[1] Country	Population (millions) 2013	Pre-primary	Primary	Secondary	Tertiary
Georgia	4.3	58% ('08)	NA	101% ('13)	33%
Iran	77.4	38% ('13)	119%	86% ('12)	58% ('13)
Iraq	33.8	7% ('07)	107%	53%	16% ('05)
Israel	7.7	112% (‘13)	104%	102%	67%
Jordan	7.3	34% (‘12)	98%	88%	47%
Kuwait	3.4	81% (‘07)	106%	100%	28% (’13)
Lebanon	4.8	102% (‘13)	113%	75%	48%
Oman	3.6	52% (‘13)	113%	91%	28% (’11)
Palestine	4.3	78% (‘13)	95%	82%	46%
Qatar	2.2	58% (‘13)	103% (’05)	112% (’11)	14% (’13)
Saudi Arabia	28.3	17% ('14)	NA	124%	58% ('13)

(continued)

Table 3.3 (continued)

Gross Enrollment Ratio: South Asia[1] Country	Population (millions) 2013	Pre-primary	Primary	Secondary	Tertiary
Syria	21.9	6% ('13)	74%	48%	31%
Turkey	74.0	28% ('13)	109%	102%	79%
UAE	9.3	79% ('11)	108% ('12)	84% ('99)	NA
Yemen	24.4	1% ('13)	101%	49%	10% ('11)
Kazakhstan	16.4	58% ('13)	106%	101%	55%
Kyrgyzstan	5.5	25% ('12)	109% ('13)	88%	48%
Tajikistan	8.2	9% ('11)	96% ('14)	87% ('12)	24% ('14)
Turkmenistan	5.2	63% ('14)	89%	85%	8%
Uzbekistan	28.9	25% ('11)	93%	105%	9%

Source Wikipedia, Education in Asia, https://en.wikipedia.org/wiki/Education_in_Asia

References

Kaledzi, I. (2022). *Why education remains a challenge in Africa*. January 24. https://www.dw.com/en/africa-right-to-education-remains-a-challenge/a-60518000

Laas, M. (2023), The UK education system. *Expatica*, 4 August. https://www.expatica.com/uk/education/children-education/the-uk-education-system-106601/

OECD. (2016). *PISA 2015 results (Volume II): Policies and practices for successful schools*. PISA, OECD Publishing. https://doi.org/10.1787/9789264267510-en

OECD, (2022). *Education at a Glance 2022: OECD indicators*. OECD Publishing. https://doi.org/10.1787/3197152b-en

Reimers, F. (1999). Education and poverty in Latin America. Can schools make any difference? *ReVista*, Spring.

Roser M., & Ortiz-Ospina, E. (2013). *Primary and secondary education.* Published online at OurWorldInData.org. https://ourworldindata.org/primary-and-secondary-education
UNESCO, CEPAL, & UNICEF. (2022). *Regional monitoring report SDG4-E2030, education in Latin America and the Caribbean at a Crossroads, 8th September.*
United Nations (2015). *Sustainable development goals.* https://sdgs.un.org/goals
World Bank Annual Report (2017). https://thedocs.worldbank.org/en/doc/908481507403754670-0330212017/original/AnnualReport2017WBG.pdf

Chapter 4
European Education Landscape

This chapter deals with education in the EU and tries to present the many differences between various groups of countries (Sect. 4.1), which can be usefully grouped from a geographic point of view (Sect. 4.2).

Differences refer: first, to various aspect of the models of primary and secondary education (Sect. 4.1.1); the different articulation of courses from elementary school to tertiary education (Sect. 4.1.2); various features of education in the EU, such as the length of school in primary and secondary schools in terms of number of days per year and the number of hours per day; the Erasmus program; whether education is free or not; education by sex; the change in the education levels of the population (Sect. 4.1.3).

Regional focus (Sect. 4.2) shows that the main systems are: the Scandinavian one; that of the Central European countries and the one of Southern European countries and France; each one of these systems shares many common features.

4.1 Schools in the EU: Differences Between Groups of Countries

4.1.1 Models of Primary and Secondary Education

Three main organisational models can be identified for all European compulsory education systems. These are: the Single structure education, the Common core curriculum provision and the Differentiated lower secondary education.

With the first model all students follow a common curriculum providing general education from the beginning to the end of compulsory education; in addition, there is no transition between primary and lower secondary education.

N. Acocella, *Understanding Education in the European Union*,
SpringerBriefs in Education, https://doi.org/10.1007/978-3-031-62065-2_4

With the Common core curriculum provision, all students progress to ISCED level 2,[1] i.e., the lower secondary level, following a common core curriculum, after successfully completing primary education, i.e., ISCED level 1.[2]

With Differentiated lower secondary education students follow distinct educational pathways or specific types of education, after successfully completing primary education. The distinction can start at the beginning or in the course of lower secondary education. A different certificate is then given to the students at the end of their studies.

Looking at Fig. 4.1, we can say that the Single structure education is common to all Balkanic states (except Greece), Denmark, Estonia, Iceland, Poland, Portugal, Scandinavian countries; the Common core curriculum provision is typical of Belgium, France, Greece, Ireland, Italy, Romania, Spain, Turkey. Finally, Differentiated lower secondary education is typical of Austria, Germany, Lithuania, Luxembourg, Netherlands, Switzerland.

In Czechia, Latvia, Hungary and Slovakia, compulsory education is of the Single structure kind up to the age of 14, 15 or 16, according to the country, but from the age of 10, 11 and 13 (depending on the country), students can be enrolled in separate educational institutions providing lower and upper secondary education (European Commission, EACEA, Eurydice, 2022a).

4.1.2 *Pyramid of Courses from Elementary Schools to Universities*

As for other countries, in the EU learning passes through different steps, from elementary schools (really, we could say that also pre-primary learning is relevant) to various grades—or cycles—of university (the Bachelor's degree, lasting three to five years, granting access to the Master's degree, normally lasting two years, and finally the Ph.D. Programme, taking four to six years). The Learning Pyramid has several different names such as a Learning Cone, Pyramid of Retention, Cone of Learning, Pyramid of Learning and even Cone of Experience.

The first time this concept appeared was in 1954 in a book called the Audio-Visual Methods in Teaching, expounding on the model that was first developed by the National Training Laboratories in Bethel, Maine.

The learning pyramid describes various learning/teaching methodologies and the retention rates of information from these methodologies.

---

[1] The International Standard Classification of Education (ISCED) is due to the United Nations. It classifies education programmes by their content using two main cross-classification variables: levels of education and fields of education. The levels established by the 2011 classification go from the level 0, ISCED 0 (for early childhood education, 'less than primary') to level 1 (Primary education), ISCED 2 (Lower secondary education), ISCED 3 (Upper secondary education), and so forth, up to ISCED 8 (Doctoral or equivalent level).

[2] This model is implemented also in the United States.

Main models of primary and lower secondary education (ISCED 1-2) in Europe, 2022/2023

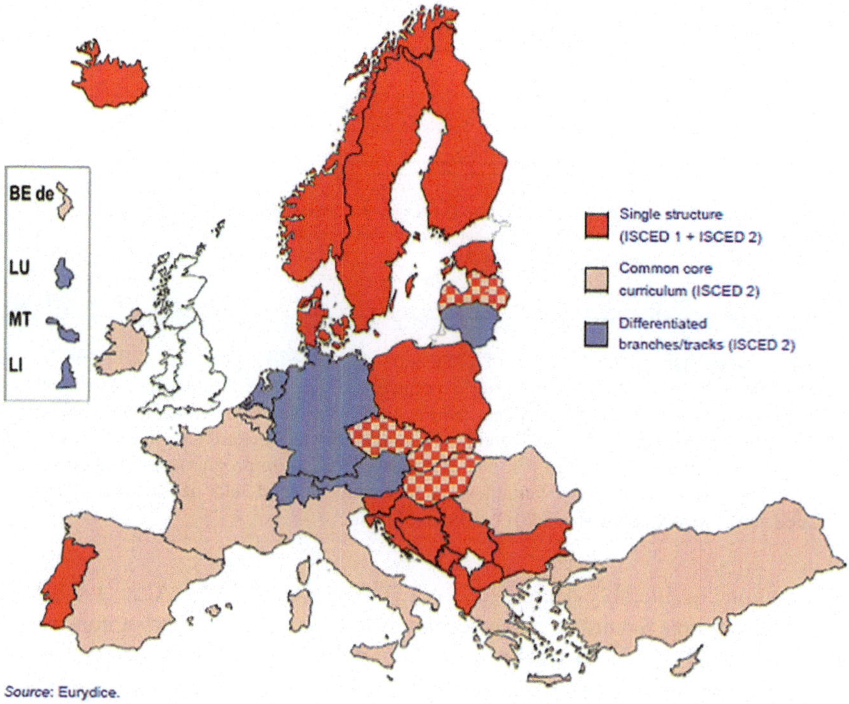

Source: Eurydice.

Note: in Czechia, Latvia, Hungary and Slovakia, compulsory education is organised in a single structure up to the age of 14, 15, and 16 depending on the country. However, from the age of 10, 11 and 13 (depending on the country), students can enrol in separate educational institutions providing lower and upper secondary education.

Fig. 4.1 Models of primary and lower secondary education in Europe, 2022–2023. *Source* European Commission, EACEA, Eurydice (2022a)

4.1.3 Various Features of Education in the EU

These are: (a) the change in the education levels of the population; (b) length of schooling in primary and secondary education (number of days per year); (c) length of schooling in primary and secondary education (number of hours per day); (d) Erasmus, the program for the mobility of university students, and other European initiatives for education; (e) is education free in Europe? (f) education by sex.

Let us deal with each one of these features in turn.

(a) the change in the education levels of the population; b) the length of schooling in primary and secondary education (number of days per year)

In the last decade the percentages of the population aged 25–74 having attained the various educational levels changed considerably in the EU. By looking at the following Fig. 4.2, we can see that people having at most the lower secondary level

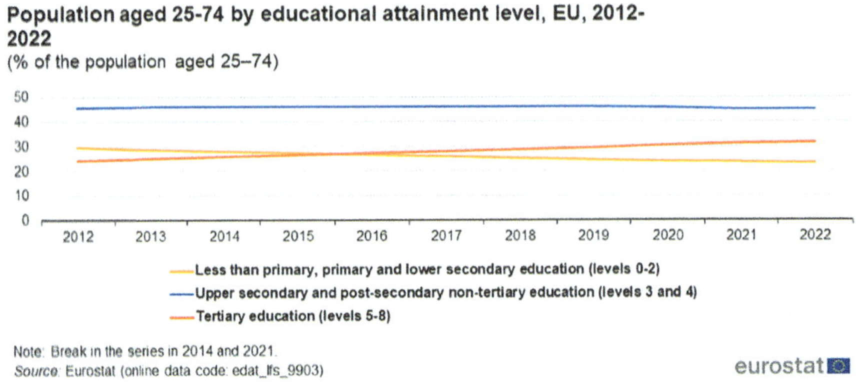

Fig. 4.2 Educational attainment level of population aged 25–74 in the EU, 2012–2022. *Note* Starting from the top in 2012, curves refer to: upper secondary and post-secondary non-tertiary (ISCED levels 3, 4 and 5), less than primary, primary and lower secondary education (ISCED levels 0, 1, 2), tertiary education (ISCED 6, 7 and 8). The International Standard Classification of Education (ISCED) is due to the United Nations. It classifies education programmes by their content using two main cross-classification variables: levels of education and fields of education. The levels indicated above were established in 2011. *Source* Eurostat

of education considerably decreased (from 30% to around 23%). Almost similarly, the share of people having an upper secondary and post-secondary (but non-tertiary education) level decreased feebly, from around 47% to 45%. By contrast, people having a tertiary education level increased from 24 to 31%. Thus, the average education level has sensibly increased in the EU. However, one has to note that, in contrast to this, the number of people not in education nor in employment or training (the so-called NEET) has increased at a fairly high rate, especially in some countries (see below, Sect. 5.2).

Extending the analysis to OECD countries, we see that, with the exception of Ireland and Luxembourg, European countries are not at the top of the percentage of people having tertiary education, where are, instead, Korea and Japan. By looking at the different percentages of people having different ages in these countries one can also see that education in countries such as Korea, Portugal and Turkey is rather recent (see Fig. 4.3). In fact, a much lower percentage of older people having this education indicates its rather recent diffusion eliminate: of it in the country. Typical are the cases of Turkey, Portugal and Korea, where the differences between the two percentages are higher than half of the values referring to the younger population.

(b) Length of Schooling in Primary and Secondary Education (Number of Days per Year)

In coming across the different countries, we note practically no difference between the different systems in some characteristics of teaching—e.g., in the length of the primary and secondary education—in terms of the total number of days and of the length of summer holidays (see Fig. 4.4 and 4.5). However, we must say that the information derivable from both figures is not completely reliable. In fact, contrary

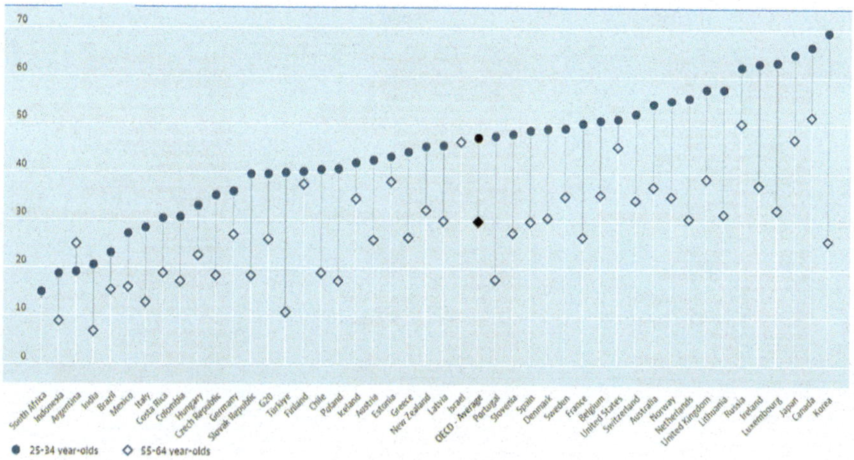

Fig. 4.3 Population of different ages with tertiary education, OECD countries, 2021 or latest available. *Source* OECD (2022)

to what derives from Fig. 4.4, the total number of schooldays is equal in France and Italy, because, with reference to Fig. 4.5 the length of summer holidays only is not reliable, without considering also that of holidays in other periods of the year, e.g., those for Christmas, the New Year and Easter. Taking considerations of both into account one could say that French and Italian elementary and secondary schools have the same length.

In fact, if we look at the following Fig. 4.6, we have a satisfactory explanation, as, e. g, for French elementary schools there are holidays for almost 70 days during the year, apart from summer holidays. The total number of holidays in this country is thus (almost) equal to those of Italy.

(c) Length of Schooling in Primary and Secondary Education (Number of Hours per Day)

A further indication of school duties, in addition to the number of days of education, is offered by the hours of compulsory education per day, which is shown by Fig. 4.7.

If we consider only the total number of hours for primary and lower secondary education, we note that below the OECD average are most of the former Soviet bloc countries, with Poland having the least number of hours (a bit higher than 5), Finland, Austria, Sweden, Belgium, Germany and Italy, whereas other European countries have more than the average, with Denmark having the highest number (more than 10).

As a consequence of the number of learning days of school and learning hours per day—in addition to other factors, such as attendance and effectiveness of teaching—we have different educational attainment levels in the various countries, as shown by Table 4.1. Obviously, the educational attainment levels of the population in the various countries are a consequence of the educational policies of different historical

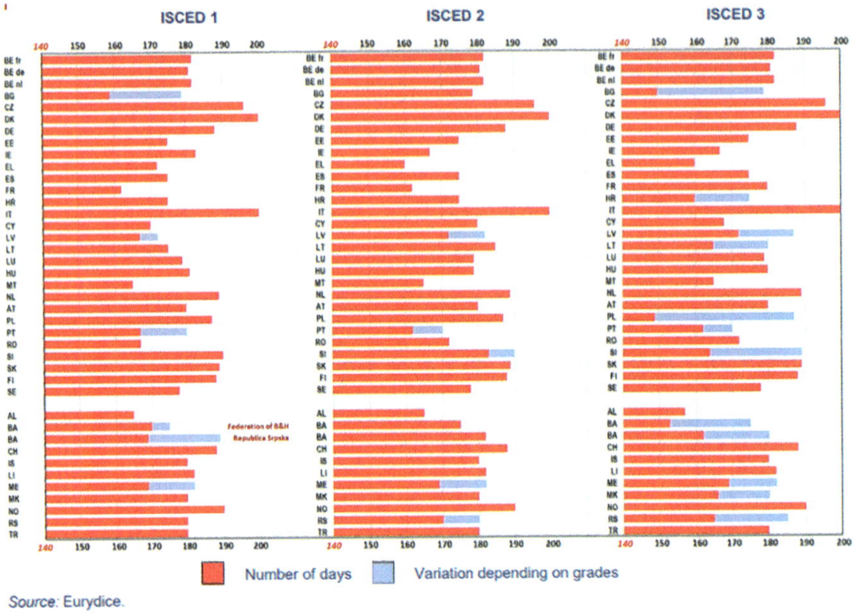

Fig. 4.4 Length of school, primary and secondary education, number of days, 2021/2022. *Source* European Commission/EACEA/Eurydice (2021)

periods. This is reflected in the Table 4.1, which offers information for the attainment levels of people having different ages, which, to a large extent, is an effect of the policies prevailing at the usual ages of education. Cases of improvement in the level of education after those ages are, in fact, infrequent.

(d) Erasmus, the Program for the Mobility of University Students, and Other European Initiatives for Education

The Erasmus (acronym of *EuRopean community Action Scheme for the Mobility of University Students*) program was instituted on June 15, 1987 and finances scholarships for studying in another EU country. It was previously restricted to applicants who had completed at least one year of tertiary-level study, but it is now also available to secondary school students. Students joining the Erasmus Programme study at least three months or do an internship for at least 2 months to an academic year in another European country. The period of study abroad is recognised by the students' university or school.

Since 1st January, 2014 two other existing programs (*Lifelong Learning Programme* and *Youth in action*) added to the existing Erasmus, to have the Erasmus + Programme. This has been renewed for the period 2021–2027 with a budget almost doubled with respect to the Programme 2014–2020.

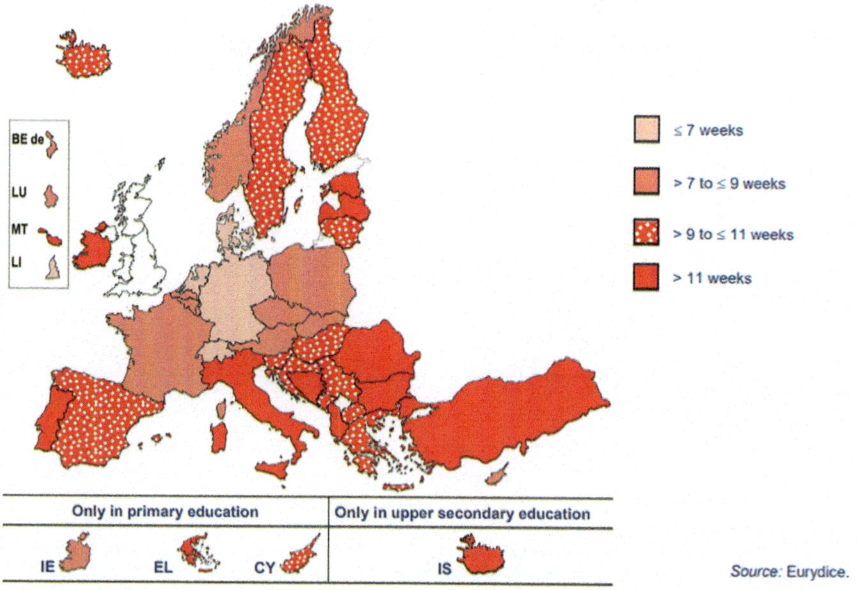

Fig. 4.5 Length of summer holidays (in weeks), primary and secondary general education, 2021/22. *Note* The length of summer holiday is: less or equal than 7 weeks in The Netherlands, Denmark, Germany, Ireland, Switzerland; from 7 to 9 weeks in Austria, Belgium, Cyprus, Czech Republic, France, Luxembourg, Hungary, Poland; from 9 to 11 weeks: Croatia, Greece, Finland, Lithuania, Iceland, North Macedonia, Montenegro, Serbia, Slovenia, Sweden; higher than 11 weeks: Albania, Bosnia and Herzegovina, Bulgaria, Estonia, Italy, Latvia, Portugal, Romania, Turkey. *Source* European Commission/EACEA/Eurydice (2021)

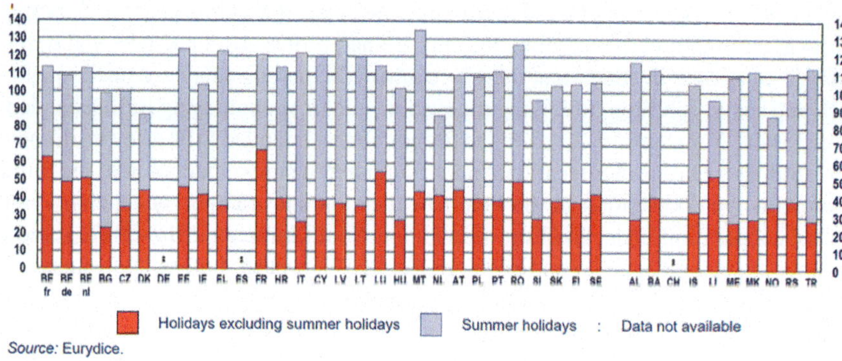

Fig. 4.6 Minimum number of holidays (in days) during the school year in primary education (ISCED 1), 2022/2023, various countries. *Source* European Commission, EACEA, Eurydice (2022b)

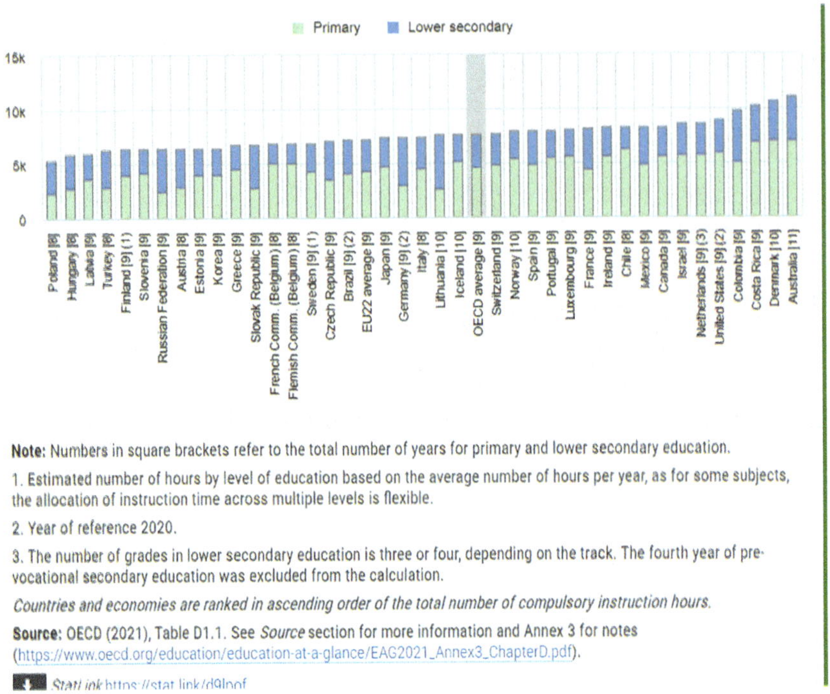

Fig. 4.7 Compulsory education hours per day in general education in OECD countries, 2021. *Source* OECD (2021)

The general objective of this Programme is to support training, youth and sport in Europe, through lifelong learning, the educational, professional and personal development of people in education, thereby contributing to sustainable growth, quality jobs and social cohesion, to driving innovation, and to strengthening European identity and active citizenship. As such, the Programme is intended to be a key instrument for building a European Education Area. It supports the implementation of the European strategic cooperation in the field of education and training, advancing youth policy cooperation under the European Union Youth Strategy 2019–2027. It has a number of sectoral agendas and develops also the European dimension in sport.

More specifically, the Programme promotes learning mobility of individuals and groups as well as cooperation, quality, inclusion and equity, excellence, creativity and innovation between organisations and policies in the field of education and training. It also promotes non-formal and informal learning mobility and active participation among young people, as well as cooperation, quality, inclusion, creativity and innovation at the level of organisations and policies in the field of youth. In addition, Erasmus supports learning mobility of sport staff, as well as cooperation, quality, inclusion, creativity and innovation at the level of not-for-profit sport organisations and sport policies (Erasmus + EU programme for education, training, youth and

Table 4.1 Population by educational attainment level for different age groups, %, 2022. *Source* Eurostat (2023)

Population by educational attainment level and Selected age groups. 2022 (% of the respective age population)										
25–54 years						55–74 years				
	Low IISCED 0–2)	Medium (ISCED 3–4)	of which		High (ISCED 5–8)	Low (ISCED 0–2)	Medium (ISCED 3–4)	of which		High (ISCED 5–8)
			General orientation	Vocational orientation				General orientation	Vocational orientation	
EU	18.1	44.2	10.2	34.1	37.7	31.6	45.7	8.7	37.0	22.7
Belgium	14.5	36.5	10.8	25.7	49.0	31.3	36.1	13.0	23.1	32.6
Bulgaria	15.8	52.6	24.7	27.9	31.6	20.5	57.8	24.1	33.7	21.7
Czechia	5.4	65.7	5.5	60.3	28.8	9.0	75.1	4.9	70.3	15.9
Denmark	16.5	37.4	7.9	29.5	46.1	25.9	44.3	4.4	39.8	29.8
Germany	17.0	49.0	5.4	43.6	34.1	17.1	55.2	2.2	53.0	27.7
Estonia	11.3	45.9	17.0	28.9	42.8	10.7	51.0	19.4	31.6	38.3
Ireland	8.8	33.3	16.9	14.8	57.9	31.0	35.9	20.7	13.4	33.1
Greece	14.5	47.9	27.1	20.9	37.6	41.4	34.4	25.2	9.2	24.3
Spain	32.1	23.1	12.8	10.3	44.8	53.7	19.8	12.6	7.2	26.5
France	13.7	39.8	9.3	30.6	46.5	30.9	43.5	10.7	32.8	25.6
Croatia	8.6	62.7	3.5	59.1	28.7	25.0	58.0	3.3	54.7	17.0
Italy	31.7	44.9	10.9	34.0	23.4	56.2	32.5	8.1	24.4	11.3
Cyprus	12.2	35.0	23.4	11.6	52.9	32.3	39.4	28.9	10.4	28.3
Latvia	9.4	47.8	20.9	27.0	42.8	6.5	64.6	22.0	42.6	28.9
Lithuania	6.5	42.0	16.5	25.5	51.5	3.5	64.3	22.4	41.9	32.2
Luxembourg	15.7	27.4		27.4	56.9	31.0	36.2		36.2	32.8
Hungary	12.5	56.1	13.0	43.1	31.4	17.4	61.3	9.6	51.7	21.3

(continued)

Table 4.1 (continued)

Population by educational attainment level and Selected age groups. 2022 (% of the respective age population)										
25–54 years						55–74 years				
	Low IISCED 0–2)	Medium (ISCED 3–4)	of which		High (ISCED 5–8)	Low (ISCED 0–2)	Medium (ISCED 3–4)	of which		High (ISCED 5–8)
			General orientation	Vocational orientation				General orientation	Vocational orientation	
Malta	29.1	35.8	19.9	15.9	35.1	65.8	21.0	11.7	9.3	13.2
Netherlands	14.7	35.8	6.7	29.1	49.5	34.7	36.4	8.6	27.8	28.8
Austria	12.5	48.4	6.2	42.1	39.1	21.5	54.5	4.7	49.9	23.9
Poland	5.7	55.5	10.0	45.5	38.8	13.2	70.4	7.0	63.4	16.4
Portugal	31.4	32.4	20.1	12.3	36.2	68.6	14.9	12.4	2.5	16.5
Romania	17.5	60.4	8.1	52.3	22.1	29.4	62.0	6.2	55.8	8.7
Slovenia	6.7	48.7	5.2	43.5	44.5	18.6	56.7	3.2	53.5	24.8
Slovakia	5.6	62.0	4.1	57.9	32.3	11.7	70.8	4.4	66.4	17.6
Finland	9.8	45.9	8.7	37.2	44.4	21.4	44.5	3.9	40.6	34.1
Sweden	10.6	36.7	14.1	22.7	52.7	18.8	47.1	17.3	29.8	34.1
Iceland	17.8	36.6	16.1	18.7	45.6	23.3	46.2	11.7	33.5	30.5
Norway	16.1	32.4	9.4	23.0	51.5	20.5	46.0	11.2	34.8	33.5
Switzerland	12.7	39.2	6.9	32.3	48.1	18.0	49.8	6.8	43.0	32.2
Serbia	13.7	58.5	4.1	54.4	27.9	31.2	52.2	4.2	48.0	16.7

Note Medium general orientation is not available in Luxembourg due to very small sample size. *Source* Eurostat (online data code: edat lfs 990:3)
Source Eurostat (2023)

sport). Finally, the Program tends to enrich students' formation through a comparison of different education systems, offering them new perspectives and indications useful for their formation.

The Capacity Building in Higher Education action of the Erasmus program aims at supporting international cooperation projects in the field of higher education in third countries not associated to the Erasmus + programme. The final purpose is to enhance socio-economic recovery, growth and prosperity and reaction to globalisation and the recent decline in human development, fragility, and rising social, economic and environmental inequalities (European Commission, 2022).

In addition to Erasmus, other initiatives have been devised by the European Commission, in order to favour education.

Thus, in 2022 the European School Education Platform was launched, due to serve as a multilingual meeting point for all those who are interested in the school education sector—either school staff, researchers, policymakers, or other professionals—at its various levels. It is free of charge for all users and tends to inform, find resources, offer opportunities for a professional development, to collaborate and develop projects, etc. Funding is provided by Erasmus + . This also funds eTwinning, launched in 2005, the community for schools in Europe, which is open exclusively to educators and other school staff in initial vocational education.

(e) Is Education Free in Europe?

Many countries offer free (or almost free, when some minimal fees have to be paid) education, both to national and foreign students, even if all students must cover their living costs and the latter have to pay travel expenses and administrative fees of few hundred euros. Students can be eligible to receive a scholarship or stipend. Many courses are offered in English. The countries which certainly follow these rules are Norway, Sweden, Denmark, Finland, Germany.

In Norway, high-quality education for home or foreign students is offered at little to no cost, as the Norwegian government finances education with taxpayers' money. However, living expenses are high in Norway.

Until 2010, in Sweden no tuition fee was paid, either by national or foreign students. As in Norway, the Swedish taxpayers would foot the bill. After 2010, tuition and application fees for non-EU/EEA[3] students have to be paid, even if scholarship programs are offered for tuition waivers, etc..

In most Danish colleges free education is offered not only for citizens, but also for those coming from other countries of the European Union, the Eurasian Economic Union, Switzerland or those having certain visa types. Some of the Danish universities (such as the University of Copenhagen and the University of Kiel) are among the best in Europe. All Danish citizens are also offered tuition aid/scholarships.

Currently, there is no tuition fee for European students attending courses in Finland. From 2017 non-EU/EEA students will have to pay tuition fees (for courses offered in English). Each University will have its fee structure. Anyway, students

[3] EEA is the acronym of European Education Area.

need to cover their living expenses. The University of Helsinki and the University of Turku are the top ranked universities of Finland.

German universities have only recently started charging tuition fees for undergraduate study programs. Right now, just 4 out of 16 Federal States—Baden-Wurttemberg, Bavaria, Hamburg, and Lower Saxony—charge tuition fees that are as little as 500 Euros per semester. All other federal states just request a small semester contribution of around 50 euros with no tuition fees charged. Tuition fees are charged for education in Austria, but they are quite low, almost insignificant for EU/EEA nationals. For non-European countries they are lower than 1000 euros per semester.

Education in Greece is also free for most programs for Eurasian nationals, but international tuition fees are quite low in comparison with many other countries (such as United States, Canada, United Kingdom and Australia).

Tuition fees for education in France are quite low, a few hundred euros for French nationals or people coming from an EU country. Fees for other international students are quite low as compared to the United Kingdom, United States, Canada or Australia, being around 3000–4000 euros per year.

(f) Education by Sex

When looking at the level of education according to the sex at the different stages, there are hardly any differences between women and men in the EU if we look at the lower education level. However, for medium and high levels different patterns can be seen.

In fact, as shown by Fig. 4.8, almost equal shares of women and men aged 25 to 64 in the EU (20.1% of women and 21.6% of men) had a low education level (lower secondary education at most) in 2021.

A smaller share of women (43.7%) than men (47.3%) in the EU had successfully completed at most the medium level of education (upper secondary education or post-secondary non-tertiary education). This pattern could be observed in almost all Member States.

When it comes to tertiary education, 36.2% of women in the EU had completed this level, compared with 31% of men. The share of women with tertiary education was higher than the corresponding share of men in almost all Member States, with the largest differences between women and men observed in the Baltic Member States as well as Finland, Sweden and Slovenia. In contrast, Germany was the only country where more men than women had a tertiary education (33% of men compared with 29% of women).

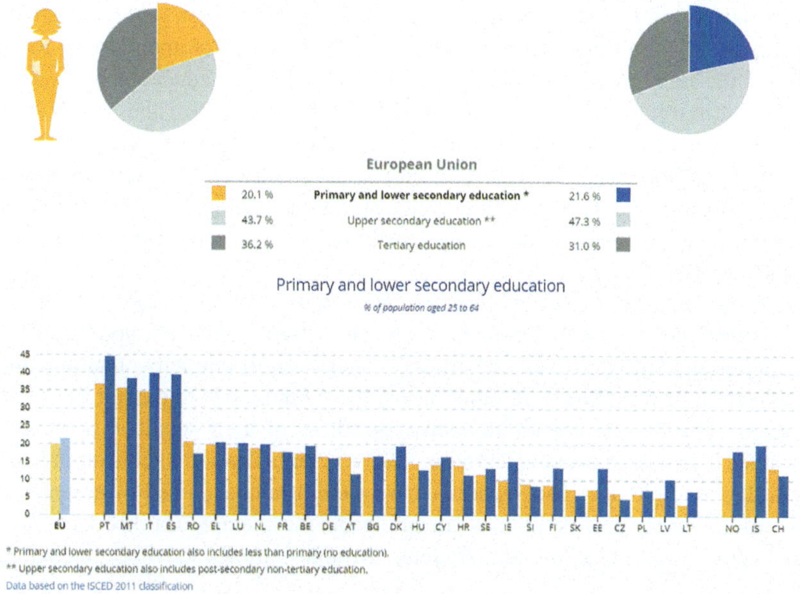

Fig. 4.8 Education levels in Europe. Primary and lower secondary education, 2021, *Source* Eurostat, https://ec.europa.eu/eurostat/cache/infographs/womenmen/bloc-2a.html?lang=en

4.2 Regional Focus

4.2.1 The Scandinavian System

The Scandinavian model of education applies to Denmark, Finland, Iceland, Norway and Sweden, i.e., also to countries that are not member of the European Union, such as Iceland and Norway. There are certainly differences as between these countries, but similarities are dominant. The distinguishing features of the Scandinavian education model—reflecting those of the whole society—are still now equity, participation, the welfare state for all and narrowing of the gaps in income in the whole society and between men and women. The period from 1945 until about 1970 is often referred to as the golden era of social democracy, and Sweden was often the main source of inspiration for reforms in the education system, particularly for Norway (Telhaug et al., 2006). Things changed starting from the 1980s, directing towards a larger role of the market in education, also in Scandinavian countries. Parents began choice of schools for their children, opting out from the control by the local authority.

The bases of the model, however, still survive. Its cornerstones are citizens' equal rights, responsibility of the state for welfare of all citizens, and struggling towards narrowing the gaps in income and between men and women.

The model has been supported by the labour market model, with collective bargaining, and a developed legislation in co-operation between governments and labour organisations. Common to these countries, as well as to others (as seen before) is the Single structure model (see sub-Sect. 4.1.1). Another commonality of these countries is the very low number of young people not attending education and unemployed or not in training (the so-called NEET). Think that in 2021 the NEET were 10.2% in Norway, as against 31.2% in Italy (see Table 5.1 in Sect. 5.2.1).

According to Moos et al. (2004), the Scandinavian countries have a strong tradition of emphasizing the role of educational institutions in the making of civic society. For these countries the schools should prepare children to have a constructive role in a democratic society.

As a consequence, the Nordic countries are often praised for leading the way towards better and more equal education, which is free, publicly funded also for higher studies and available to all, regardless of social origin, ethnicity or nationality, religious belief, or disabilities. In other terms, the model tends to pursuit the same educational opportunities for all students, regardless of their social or economic background. It promotes lifelong learning, involving organisation of education from early childhood to old age. The Scandinavian system is characterised by absence of formal evaluations for the first years of education and the attempt to build cooperative relationships between teachers and students through various activities, including games.

As a consequence, education has increased in these countries. 23% of Swedes had attained a higher education of three years or more in 2021. Women are particularly educated. As an example, in Denmark, more women than men had completed a master's degree in 2021.

Education generally follows the same trajectory in the five Nordic countries. Primary school is mandatory for nine or ten years and is followed by optional upper secondary and higher education. In order to increase female employment a strong emphasis is put on pre-primary education. Most schools and education institutions are public, and the vast majority of pupils and students attend public schools, even if in Norway and Sweden the number of private schools has increased over the past years. Despite these commonalities, however, the education systems are not identical. For example, Sweden offers one year of optional so-called preschool class the year before children enter primary education. In Norway, elementary school is divided into several stages. The Scandinavian countries also follow different national curriculums, which cannot be deviated from.

Secondary education is optional, but most Scandinavian students attend it—also in order to prepare for further higher education and university—even if the workload and the knowledge and skills requirements of children increase considerably.

Immigrants' share in upper secondary education has also increased, especially in Sweden. Some students of Nordic countries decide to study abroad in another Scandinavian country or in the United Kingdom and other countries.

English is usually taught in schools from the age of 7 or 8 onwards in all Scandinavian countries and is treated like other core subjects, such as mathematics.

Special care is put in the selection of teachers. As an example, in 2011, the Swedish Education Act required for them higher formal requirements and professional certification than earlier.

Scandinavian universities are of good quality and at least three of them are listed among the first one hundred, according to the Times Higher Education World University Rankings.

Frønes et al. (2020) say that the Scandinavian model aims at offering equal opportunities and inclusion to all the students and is based on the concept of 'Education for All', including the marginalised students. It does so by removing barriers to educational opportunities, also for students with special needs.

According to Beach (2018), instead, the Swedish education model shows serious structural inequalities as to race, class, gender and special needs.

From the 1930s to the 1980s the Folk Home educational reforms were introduced. They tended to create a more inclusive school offering equal educational possibilities sustained by high investment. However, more recently the reform process—which had never had any particularly egalitarian effects—has been abandoned. In fact, the neo-liberal paradigm has been adopted by the new right-wing government and differences between social classes and groups in terms of family income, parental occupation and education have increased and widened at each stage of the educational system. In other terms, the shortcomings of the Swedish system have increased recently, due to marginalisation and the segregation of socially disadvantaged and ethnic minority groups. Therefore, differences in education and performance levels have widened, and the rhetoric of inclusion has become a metaphor for the dominance of human capital, manifested in personal choice over social justice. Sweden's education system is perhaps internationally seen as remarkable, inclusive and access based. However, perhaps equally remarkable is the lack of impact deriving from investment in terms of significantly reducing gender, racial and ethnic inequality or social differences.

Statistics—such as those behind the Table 4.1 in sub-Sect. 4.1.3—say clearly that not only other Scandinavian countries, but also Sweden, are at the top of education. However, we must say also that data in Henrekson, Wennström (2022) show that Sweden's results steadily deteriorated in the years to 2012 in all three PISA areas—reading, mathematics, and science—and Swedish average score was well below the OECD average. Starting in 2015, Sweden has recovered a large part of the previous decline, but this is to some extent due to the deterioration of the OECD average scores and possibly to other factors, partly related to this.

4.2.2 Central European Countries

Contrary to what happens in Scandinavia, different models of education are implemented in the different Central European countries. Common to Austria, Germany, Luxembourg, the Netherlands and Switzerland is the model of differentiated education. In the Czech Republic, Latvia, Hungary and Slovakia, there is a mixed system, of single structure and differentiated education (see Sect. 4.1.1). Poland (together with most Balkan states) has a single structure education.

By contrast to other European countries, education in Germany is primarily the responsibility of individual German states (*Länder*) and the federal government plays a minor role. Optional pre-primary school is provided for all children from one to six years, but after this age attendance is compulsory. In recent years the performance of German students has been above the OECD averages. More in detail, according to PISA 2022, students in Germany scored overall rather high in reading (480 points), mathematics (475) and science (492), against the OECD average of 476, 472 and 485, respectively. To make a more detailed comparison, their performance was well above that of the Americans, but in 2022 the average performance in reading fell to very low levels, below those of 2012. Top performers were, however, not uncommon, offering thus a high-educated labour force. Secondary Germany education schools are: the *Gymnasium*, the grammar school for high-performing students, preparing them for higher education; the *Realschule* for middle-performing students, who study until 16 and then either attend *Gymnasium* or continue for a vocationaldiploma; the *Hauptschule* for students up to the age of 15, after which they can attend an extra full year at the *Realschule* or attend training programmes usually lasting between one and three-and-a-half years, which are either school-based or combine classes with practical training in an establishment or company; the *Gesamtschule*, which integrates *Realschule* and *Hauptschule*. Students having a qualification or a *Meisterbrief* (master craftsman's diploma) granted at the end of secondary education can attend German universities. For universities of applied sciences also other diplomas are accepted.

In Austria, primary education is obligatory since 1774, due to Empress Maria Theresa. Now it is mostly provided in public schools (about 90% of all schools) and is compulsory for all children aged 6 to 12, first in elementary school (*Volksschule*) and for other 4 years in a lower secondary education (*Mittelschule*) or a grammar school (*Gymnasium*). Separate routes are then open to those who want to take up an apprenticeship—who attend a polytechnic institute (*Polytechnische Schule*) for one year and then find apprentice position and attend vocational school (*Berufsschule*) for three years—and children who attend an institution of higher education for four to five years, usually after an entrance exam. After a final exam, students can go to university.

In Switzerland the education system is devolved primarily to the cantons, but primary school is mandatory for every child—usually from the age of 6—and is free (at least in state schools). All children are required to attend school for at least 9 years. No many students (only about 25%) attend both lower and upper secondary schools. The other students follow either theoretical or practical courses. Many university students come from abroad.

In the Netherlands education is mostly public and is government-financed, but there are also a few private schools. The Dutch grading scale runs from 1 (very poor) to 10 (outstanding). In elementary and high schools, pupils are assessed annually by a team of teachers. Compulsory education starts at the age of four-five and ends at 15–16; afterwards, children must attend at least a partial education up to 18. The Dutch system has some specificities, e. g., in not having middle schools or junior high schools and in not only moving up pupils, but also in demoting them to a lower level of education, when they cannot cope with the topic of the current level, or when they have poor grades. Secondary education in the Netherlands has three programs: pre-vocational training, followed by vocational training; senior general, followed by higher professional training; pre-university, followed by university education. Attending one or another program depends—among the various factors—on the pupil's performance in the primary school final test. For graduating from a Dutch secondary school, students must pass both school exams (at mid-term) and national exams (at the end of the year). Students who fail an exam by a small margin can retake it. If failure is by a wide margin across multiple subjects, graduation is not allowed and the students must redo their final year curriculum. PISA ranks now Netherlands' education as outstanding, the 16th best in the world as of 2018, being significantly higher than the OECD average, but with a declining position with respect to previous tests. The NEET ratio in this country (4.5%) was the lowest among OECD countries in 2021 (see Table 5.6). The Dutch education system is rather variegated between cities and is very different from most other countries, but offers many choices and works very well. The state funds equally both schools following particular religious or pedagogic principles and public schools since 1917.

In the case of the Netherlands, socio-economic status does not have a significant impact on students' participation in education. In addition, in 2015 private sources accounted for only 13% of total expenditure in pre-primary institutions (less than the OECD average of 17%), even if at the tertiary level the Dutch private percentage (29%) was closer to that of the OECD countries (30%).

Large differences in educational attainment may lead to starker earnings inequality in many countries. In the Netherlands, 32% of 25–64 year-old adults with below upper secondary attainment had earnings lower than or equal to the median earnings in 2019, above the OECD average of 27%.

Education in Luxembourg is multilingual. It starts with fundamental education. Secondary education and higher education follow. Most schools are public and free. Attending school is obligatory from the ages of 4 to 16. Fundamental education is given by: preschool and primary school (from the age of 3 to 11); secondary education (lasting 6–7 years, articulated in: classical system; technical system; general regime; technician training regime; professional regime; finally, there is a preparatory regime, for those who have not fulfilled the requirements of fundamental education).

Education in the Czech Republic is articulated in preschools (for pupils aged 3–5), elementary school (from 6 to 15), consisting of 2 stages, the first one for 5 years and the second for 4 years; secondary school (formal education; professional or vocational school), and post-secondary school. All schools, except for preschools, are state funded.

In Slovakia most schools, especially universities, are public. There are 10 years of compulsory education (first primary education stage for pupils aged 6–10, second primary education stage for ages 10–15); high schools for ages 15–19, divided in general education, gymnasium, various vocational schools, tertiary education (3 years for a bachelor's degree and 2 years for masters). There are rather few NEET in this country.

In Hungary schools are mostly public and free (including universities). They are compulsory from kindergarten (for pupils 3–6) to primary education (until the age of 16). Secondary education includes: the Gymnasium, preparing students for university studies; the secondary vocational schools for intermediate students; the technical school preparing for work. TIMSS, i.e. Trends in International Mathematics and Science Study, has rated 13–14-year-old pupils in Hungary among the best in the world for mathematics and science.

Education in Latvia is mostly public and free. It is compulsory (from preschool education, for ages 5 to 7, to primary education for additional 9 years). In the last two decades the share of 25–34 years-old having a tertiary qualification has increased by 28 p. p., reaching levels very similar to the 47% average of OECD countries. In Latvia, the share of women among general upper secondary graduates is 54% (again, very close to the OECD average of 55%), while men are 56% of all vocational upper secondary graduates.

In Poland there are eight classes as primary schools, followed by general high schools or technical and vocational schools, at the end of which students must pass the matriculation examination, giving access to universities and colleges. The performance of Polish students in PISA evaluations is very good in all three fields and well above the OECD average (see Table 5.3 in Sect. 5.2.3).

4.2.3 The Balkan States

The situation of education in the Balkan region is very complex, adding to—or deriving from the Balkan's complexity in terms of ethnicity, language and religion. It may be really the case of not to think of the Balkans as a whole, even if the Balkan countries (i.e., Albania, Bosnia and Herzegovina, Bulgaria, Croatia, Kosovo, Macedonia, Montenegro, Romania, Serbia, to which many people add Greece) are developing their collaborations on education.

The European Union has set some standard aims. Since the break-up of Yugoslavia in 1990 a number of changes have occurred. As for our field of interest, there has been a steep reduction in government funding and the partial disintegration of the education systems existing before the war, the integration of these systems into the Bologna process (see Sect. 6.8) and the European higher education area, and structural reforms in the public sector affecting education, even if corruption has a negative influence, especially in the field of higher education. Other shortcomings refer to the poor educational infrastructures, the paucity of human resources for education and the high dropout rate due to poverty and scarcity of infrastructures.

Compulsory education usually starts at the age of 6 in Western Balkans and Bulgaria, and at 4 in Romania. It is compulsory up to the age of 15 (18 for Bulgaria and 19 for Romania). The lowest literacy rate was 97.55% in Albania, below that in the other Balkan states.

Most Balkan countries—the main exception being Bulgaria, which follows the common core model of education—share the single structure model of education.

The performance of all Balkan students for all PISA tests in 2022 have been well below that of their EU and OECD fellows.

4.2.4 Southern European Countries and France

Education widely differs also in these countries.

Education in France is obligatory from the age of 3 and is divided into preschool, primary education, secondary education, and higher education. Elementary school begins at the age of 5–6 years and lasts 5 years. It has two stages (an initial one and one of skill development) that tend to give pupils elementary capacity of reading, writing and counting. After this school students attend the middle school (*college*), from the age of 11 to 14, and, after passing an exam (*le brevet*), the high school (*le licée*) or the apprentice learning center, providing vocational degrees and imparting knowledge for a specific profession (*le licée professionnel*), for the ages of 15–18. The high school has three sections: literary, socio-economic and scientific. Higher education (university) has three levels, the bachelor's degree, the Master's and the Doctoral degree. A specificity of education in France is that school attendance is not compulsory, only instruction being so. More than 80% of schools are public. French students' scores in PISA surveys declined over the years. Even so, PISA 2018 ranked French pupils right above the OECD average in reading (493 against the OECD average of 487), mathematics (495 against 489) and science (493 against 489). Both in reading and science girls performed better than boys. France is one of the OECD countries where the link between socioeconomic status and performance is the strongest. In fact, this status explains almost 1/5 of the variance in reading performance. French students felt low support and feedback from their teachers. Only 40% of students—compared to almost 50% students on average in OECD countries—reported that they received help to improve from their teachers. Finally, French students reported a scarce disciplinary climate at school.

Education in Greece is centralised. Public schools provide also supply of textbooks, are tuition-free and are funded by the government. There are three educational stages: the primary stage, for the ages 6 to 12; the secondary stage, divided into a compulsory middle school, lasting for 3 years, and a non-compulsory Lyceum, lasting 3 years (for the ages 15–18); the third stage involving higher education, which can be accessed after examinations. Private tertiary education is not allowed in Greece. Private institutes for primary and secondary education are not forbidden, but cannot be given public funding.

In Italy public education prevails, but there are also private institutions. Education is divided into five stages: kindergarten, primary school, lower secondary school, upper secondary school (lyceum with different specialisations: classical, scientific, language, human sciences, economic-social, art; technical institute; professional institute), and university, with BA, Masters and Ph. D. titles. Education is compulsory from 6 to 16 years of age—also for children of other nationalities who are resident in Italy—and free up to the university, where a fee must be paid, covering

36% (a level higher than the 30% of the OECD average) of costs, when academies are public, and equal to 100% when they are private. A fee must be paid also for access to private pre-primary institutions, accounting for 19% of total expenditure (a little higher than the OECD average of 17%). The state does not spend too much (less than other European countries) for education. PISA 2018 reported the Italian secondary education below the OECD average in reading and science, and near the OECD average in mathematics, with a wide gap between Northern schools, which performed near average, and schools in the South, with much poorer results. Part of these poor results is due to absences and indiscipline of students in classrooms. The socio-economic status may significantly affect not only students' participation (especially to academic courses, requiring payment of fees), but also their performance, more than the gender and immigrant status. In Italy, 1.9% (3.3%) of students in lower secondary (and in upper secondary initial education) repeated a grade in 2019 (whereas the OECD averages were 1.9% and 3%, respectively) and the number of boys was higher than that of girls. In 2019 among the world's top 500 universities there were 33 Italian, the third-largest number in Europe, after the United Kingdom and Germany. Education and training system dropouts have increased, reaching the level of almost 12% of young people aged 18–24 in 2022. In 2021, the share of 25–34 aged people with tertiary attainment in Italy had raised by 18 p.p. from 10% in 2000 to 28%, but was much lower than the average OECD one (47%). One of the reasons why education is in rather bad shape in this country is the relatively low expenditure: in 2019, this was 10,902 $ per full-time equivalent student in primary to tertiary educational institutions, compared to 11,990 $ on average in the OECD countries. The relatively low level of pay for educators is one of the reasons why there is a high number of job vacancies. This is also due to the skills possessed by aspirants to the role of teachers, which are lower than—or different from—those required by existing vacancies, especially in digital initiatives.

Also in Malta there are the usual five stages of education: pre-primary (for the ages 3–5), primary (ages 5–11), middle (ages 11–13), secondary (13–18) and tertiary. Education is compulsory from age 5 to 16. Pre-primary education is optional, but free. At the end of primary schools, students are subject to an examination whose results allow them to be enrolled in junior lyceums (if results are highly positive) or in a less competitive secondary school. There are 3 different providers of education: the state, the Catholic Church, and other institutions of the private sector. The state runs 70% of primary and secondary schools. State schools provide textbooks free of charge to their students.

In Portugal from the 1960s, public education is mandatory for all children between the ages of 6 and 12—but many parents enroll their children before the age of 6—and the number of public secondary schools and universities expanded. Education in Portugal is articulated into: pre-primary education (nursery, up the age of 3; kindergarten, from 3 to 5, which is provided by the state—in which case it is free—and private institutions); basic education, lasting for 9 years, divided into 3 stages, with exams on Portuguese and Mathematics at the end of the third stage; secondary education, compulsory since the school year 2012/2013 and consisting of a three-year cycle, with 3 types of programmes: general programmes, technical/vocational programmes, and artistic programmes. Higher education is articulated into universities (highly research-oriented) and polytechnics (profession-oriented and offering a more practical training), either public or private, giving BA and Master titles. Low-income families can apply for government support for their children and for a family allowance. Families of children with disabilities can also receive various kinds of financial support from the government. Some state schools offer language support classes for foreign children who don't yet speak Portuguese and their final Portuguese exams can also be adapted for non-native speakers (Expatica, 2023). Schools are public and private and are regulated by the State through the Ministry of Education. Tertiary education is overseen by the Ministry of Science, Technology, and Higher Education (*Ministerio da Ciência, Tecnologia e Ensino Superior*—MCTES). Despite the fact that Portugal has historically had a high illiteracy rate, the country's current level of education is actually relatively good. 92% of children attended preschool in Portugal in 2019, which is higher than the OECD average of 88%. The basic literacy rate of the Portuguese population is 99.44%. According to PISA 2022, Portugal's score is practically equal to the OECD's average for all proofs, whereas in 2018 it was similar to Norway, Poland, Denmark and Belgium. The first Portuguese university (Coimbra) was founded in 1290. In the centuries, the illiteracy rate rose and by the end of the nineteenth century it was over 80%. In the following years it fell down, as we have seen.

Education in Spain is compulsory and free for all children aged between 6 and 16 years and is supported by the national government together with the governments of each of the country's 17 autonomous communities. Spain ranked practically equal to the OECD PISA 2022 average for the three proofs. Primary (for children aged 6–12 years) and lower secondary schools (12–16 years old children) are mandatory. Access to lower secondary education is automatic, i.e., no exam is required, whereas a school-leaving certificate is required to access a high school (*bachillerato*), which prepares students for university or a vocational school. Students who want access to university are required to attend two additional final years in high school. There are then 3 levels of education for professional skills acquisition besides the university education. As of 2020–21, Spain has almost 10 million students (4,654,727 attending primary education and 2,730,036 in secondary schools; 1,633,358; vocational education, with 887,710 students). There are also preschool cycles for the ages of 0–3 and 3–6, free when provided by public centres. The medium of instruction is Spanish, coexisting with Basque, Catalan, Galician and Aranese, which is the primary language of a section of Catalonia. Schools in Spain are: state schools (67.4% of pupils), privately run schools, but funded by the state (26.0%), purely private schools (6.6%). All non-university state education is free in Spain.

References

Beach, D. (2018). The myth of Swedish education equity. In: D. Beach, *Structural injustices in Swedish education* (pp. 1–32). Palgrave Macmillan. https://doi.org/10.1007/978-3-319-95405-9_1

European Commission. (2022). *Erasmus + Programme Guide, Version 1 (2022): 24-11-2021.*

European Commission, EACEA, & Eurydice. (2021). *The organisation of school time in Europe. Primary and general secondary education—2021/22. Eurydice facts and figures.* Publications Office of the European Union.

European Commission, EACEA, & Eurydice. (2022a). *The structure of the European education systems, 2022/2023, Eurydice facts and figures.* Publications Office of the European Union.

European Commission, EACEA, & Eurydice. (2022b). *The Organisation of school time in Europe. Primary and general secondary education—2022/2023. Eurydice facts and figures.* Publications Office of the European Union.

Eurostat. (2023). Educational attainment statistics. https://ec.europa.eu/eurostat/statistics-explained/index.php?title=Educational_attainment_statistics

Expatica. (2023). *The education system in Portugal.* https://www.expatica.com/pt/education/children-education/the-education-system-in-portugal-105195/

Frønes T. S., Pettersen, A., Radišić, J., & Buchholtz, N. (2020). Equity, equality and diversity in the Nordic model of education. Contributions from large-scale studies. In: T. S. Frønes, A. Pettersen, J. Radišić, & N. Buchholtz (Eds.) *Equity, equality and diversity in the Nordic model of education.* Springer, Cham. https://doi.org/10.1007/978-3-030-61648-9_1

Henrekson, M., & Wennström, J. (2022). Educational performance in Swedish schools—What are the facts? In M. Henrekson, & J. Wennström (Eds.), *Dumbing down. The crisis of quality and equity in a once-great school system—and How to reverse the trend.* Palgrave Macmillan.

Moos, L., Möller, J., & Johansson, O. (2004). A Scandinavian perspective on educational leadership. *Educational Forum, 68*(3), 200–210.

OECD. (2021). Education at a glance 2021: OECD indicators. OECD Publishing. https://doi.org/10.1787/b35a14e5-en

OECD. (2022). *Education at a glance 2022: OECD indicators.* OECD Publishing. https://doi.org/10.1787/3197152b-en

Telhaug A. O., Mediås, O. A., & Aasen, P. (2006). The Nordic model in education: Education as part of the political system in the last 50 years. *Scandinavian Journal of Educational Research, 50*(3), 245–283. https://doi.org/10.1080/00313830600743274

Chapter 5
Comparative Analysis

Introduction

This chapter is divided in two main sections, dealing with: expenditures for education (Sect. 5.1) and various aspects of the performance of education in Europe (Sect. 5.2). Expenditures are met by public bodies or by families and other private sources (Sect. 5.1.1) and are different in the two systems, being often—but not always—higher when there are private providers (Sect. 5.1.2). The performance of the various systems differs as to the attendance, assessment methods, ages for completion of studies, the possibility to find a job and social attainments, the role of teachers (Sect. 5.2).

5.1 Education Expenditure

5.1.1 State and Private Expenditure

Investment in education by a country tends to help foster its economic growth, by enhancing productivity, to contribute to personal and social development, and to reduce social inequality. Obviously, the total level of expenditure for education is affected by the size of the school-age population, the age composition of the population, enrollment rates (which are close to 100% in most OECD countries at the primary and lower secondary levels), the levels of teachers' salaries, and the organisation of education systems. However, the amount devoted to education in public budgets depends on the general policies of each government, since this destination must be balanced against other items of public spending and all items depend on the phase of the economic cycle and the level of the GDP of the country.

Government education expenditures have been very different in the various world regions in 2015 and 2021. Vis-à-vis a world value of 4.4% with respect to the GDP

N. Acocella, *Understanding Education in the European Union*,
SpringerBriefs in Education, https://doi.org/10.1007/978-3-031-62065-2_5

Table 5.1 Expenditures on education and trained teachers, various world regions, 2015, 2021

Gov. Exp. (% of GDP)	Gov. Exp. (% 2015)	Cap. Exp. (2021)	Trained teach. Pre-primary (% total teachers)	Trained teach (% total teachers)	Trained teach. (% total teachers)					
World	4.4	"	13.5	"	"	"	85.8	"	"	"
East Asia and Pacific	3.8	"	15.3	14.8	"	"	"	"	"	"
Europe and Central Asia	4.9	"	11.8	"	"	"	94.1	"	"	"
Latin America and Caribbean	4.5	"	16.2	16.1	80.2	"	81.4	"	81.0	"
Middle East and North Africa	4.2	"	11.6	"	79.8	"	86.4	"	85.0	"
South Asia	3.0	"	12.8	10.6	"	"	73.8	"	"	"
Sub-Saharan Africa	3.7	"	16.7	14.3	54.7	"	66.9	"	58.7	"
Low income	3.3	"	15.5	13.2	45.2	"	70.3	"	60.2	"
Low middle income	4.1	"	15.3	15.3	"	"	80.1	"	"	"
Upper middle income	4.0	"	14.0	"	"	"	"	"	"	"
High income	4.9	"	12.3	"	"	"	"	"	"	"

Source The World Bank

(the values for this indicator are available only for 2015), there were lower values of South Asia (3.0%), not only lower than the world value, but also very different from the highest values of Europe and Central Asia (4.9%). The latter value is exactly equal to that relative to high income countries. In 2015[1] government expenditures as a share of capital expenditures were lowest in the Middle East and North Africa (11.6%), against the world value of 13.5% and much lower than the Sub-Saharian Africa (16.7%), which is really a favourable figure, indicating that the countries of this region heavily invest in their future (see Table 5.1).

Another relevant piece of information refers to education expenses for the different grades of school in the various OECD countries. Tertiary expenses were highest in Norway, Austria, India, Denmark and Sweden. In part in a similar way, the ratio of

[1] Data for the world as a whole are unavailable for 2021, but data available for all world regions are lower for this year with respect to 2015, except for lower-middle income countries, for which they are equal in the two years (see Table 5.1).

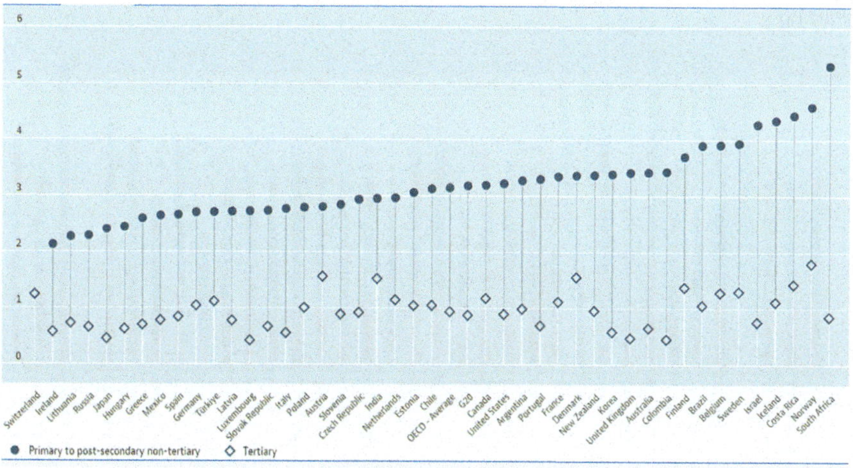

Fig. 5.1 Public spending on education in OECD countries. Primary to post-secondary non-tertiary/ Tertiary, % of GDP, 2020 or latest available. *Source* OECD, https://data.OECD.org/eduresource/public-spending-on-education.htm

expenses for primary to post-secondary education with respect to tertiary education was higher in South Africa, Norway, Costa Rica, Iceland and Israel (see Fig. 5.1).

Figure 5.2 refers more specifically to government expenditure on education as a % of GDP in the European Union. Apart from the top value of this indicator (almost 8%), referred to Norway (which is not a member of the EU), the highest value in the EU is that of Sweden, followed by Belgium, Denmark, Estonia and Slovenia, whereas Ireland and Romania have the lowest.

In OECD countries, the average expenditure per student on educational institutions from primary to tertiary levels is equal to 26% of per capita GDP. However, the value in each country is positively related to the level of its GDP per capita.

As to the different levels of education, on average across OECD countries expenditure on primary education amounts to 30% of the funding for educational institutions, while secondary education accounts for 39%. However, the share of expenditure on educational institutions is strongly influenced by the demographic composition in each country, as well as the duration of each level of education.

Another important value refers to pre-primary education expenses, which can be compared to those for tertiary schools. Figure 5.3 offers the reverse ratio of this, i.e. the ratio of tertiary to early childhood expenses to indicate the different weight placed by public institutions and families on these two stages of education. Apart from Luxembourg, which shows an exceptionally high ratio and is followed by the United States and the United Kingdom—testifying the relevance of their highest education institutions—Sweden spends very much for tertiary education, but has a low expenditure for early childhood education. For most other countries there is no much difference between the two items of expenditures and the first item of education

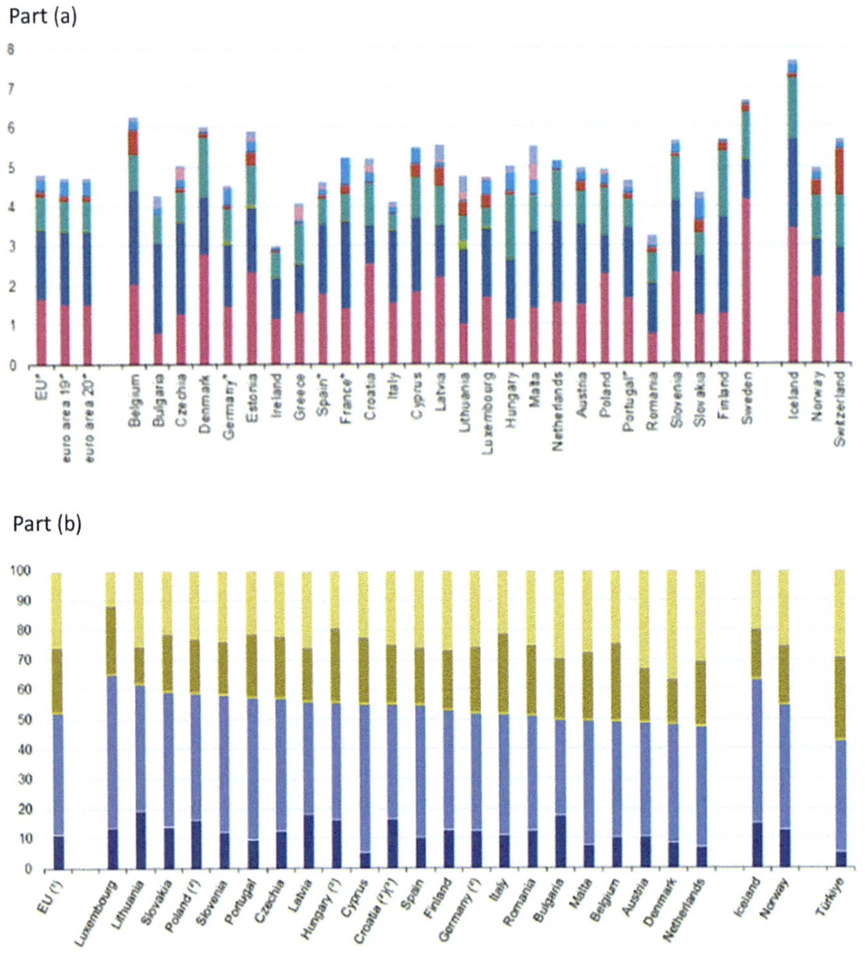

Fig. 5.2 Total general government expenditure on education in the European Union, 2021 (% of GDP) (part a) and distribution among the different levels, excluding early childhood (2020). *Source* Eurostat (2019). *Note* The histograms represent, from the bottom: pre-primary, primary and lower secondary, upper secondary and post-secondary, but not tertiary, tertiary education expenditures

expenditure—i.e., expenditure for tertiary education—is higher than the other item, with the exception of Iceland.

Some details on the data given by the previous figure as to the source of funds and the type of transaction are offered by Table 5.2. This says that government expenditures of Germany were 163 million euro, those of France around 130.5 and those of Italy about 73.5. By contrast, those of households in France were 12.4 million, against around 10 million in Italy. Germany's households spent nothing.

There are large differences between countries in the level of expenditures on education.

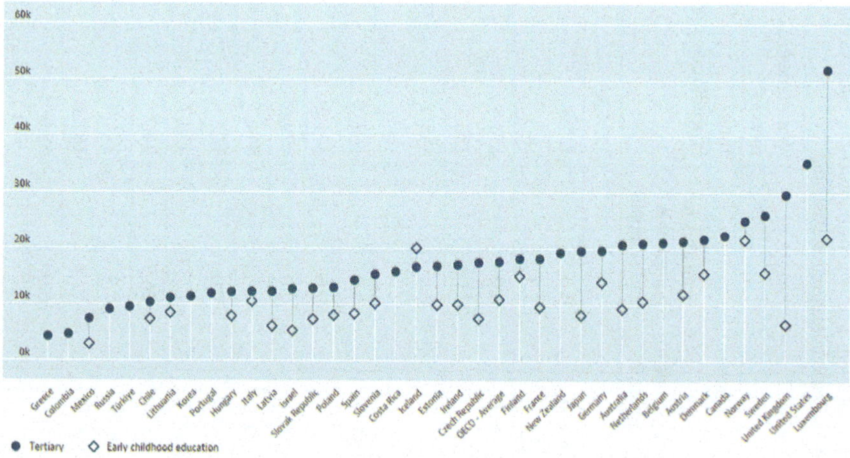

Fig. 5.3 Education spending. Tertiary/Early childhood education, US dollars/student, 2020 or latest available, OECD countries. *Source* OECD, https://data.OECD.org/eduresource/education-spending.htm

In fact, as a percentage of GDP, in the EU the highest amounts were reported by Sweden (6.7%), followed by Belgium (6.3%) and Denmark (6.0%). Among EFTA countries, Iceland reported a high ratio, equal to 7.7% GDP, followed by Switzerland (5.7%) and Norway (5.0) (Eurostat, Government expenditures on Education, https://ec.europa.eu/eurostat/statistics-explained/SEPDF/cache/42735.pdf).

At the level of the EU, general government expenditure on education as a share of total expenditure amounted to 9.4% in 2021, a stable share compared to 2020. At the country level, it was: for Estonia (14.3%), Sweden (13.5%), Cyprus, Lithuania and Malta (all 12.7%), Latvia (12.6%). For Iceland and Switzerland, education expenditure accounted for relatively high shares of general government total expenditure (15.6% in both countries). The lowest shares of education expenditure in total expenditure were observed for Greece (7.1%), Italy (7.4%), Romania (8.1%), followed by Germany and Austria (both 8.8%).

Considering the different types of transaction expenditure on education with reference to 2021, at the EU level, expenditure on education was divided into the following transactions: around 64% was 'compensation of employees' (meaning wages, salaries and employers' social contributions for, e.g., teachers); 13% was in the form of 'intermediate consumption' (meaning purchases of goods and services), 6% was in the form of 'social benefits other than social transfers in kind and social transfers in kind' (purchased market production, for example school transport) and 6% for 'other current transfers' as, for example, payments to private schools. 'Capital investments' (gross capital formation), e.g. on school buildings, accounted for around 7% of expenditure for education.

Over the period 1995–2021, at the level of the EU, government expenditure on education as a ratio to GDP increased first by 0.2 p.p., arriving to 4.8% of GDP, after

Table 5.2 Eurostat, Expenditure on education (excluding early childhood educational development), by source of funds and type of transaction, Educational expenditure statistics, 2019. *Source* Eurostat, https://ec.europa.eu/eurostat/statistics-explained/index.php?title-Educationalexpenditurestatistics

	Government			Households		Other private entities			Non-domestic sources
	Direct payments to institution	Transfers to		Total expenditure	Of which	Total expenditure	Of which		Total expenditure
		Households and students	Other private entities		Net payment to institution		Direct payments to institution	Transfers to households and students	
EU (1)	590,135.8	44,642.9	1584.9						7224.3
Belgium	28,039.6	1802.2	45.5	2559.2	1169.1	467.6	467.6	0.0	435.8
Bulgaria (2)	1854.8	419.6	0.0	482.7	319.4	9.9	5.6	3.3	58.3
Czechia (3)	8671.0	246.4	0.0	760.3	582.0	429.9	430.0	0.0	158.8
Denmark	16,645.6	3113.0	0.0	1001.2	1001.2	698.2	662.3	35.8	280.3
Germany (4)	148,895.6	14,201.4	0.0					0.0	855.4
Estonia (5)	1135.2								
Ireland (6)	8035.4		0.0	1177.4		353.4	350.8	0.0	167.7
Greece	6535.5	12.6	27.0	2329.0	784.4			0.6	374.2
Spain	48,773.5	2041.1	0.0	14,596.1	10,466.0	865.3	885.2	0.0	279.6
France	124,988.6	5631.0	1.9		12,434.7		5090.7		844.7
Croatia	2142.4	0.0	0.0	192.2	192.2	201.1	201.2	0.0	111.4
Italy	68,178.0	5358.0	43.6	11,821.6	9961.2	930.4	804.8	125.6	420.7
Cyprus	1159.7	45.5	0.0	423.6	306.1	44.5	18.0	26.5	24.7
Latvia	1316.6	37.4	2.1	246.9	157.3	66.2	56.8	9.4	121.6
Lithuania	1799.7	51.6	5.4	233.0	196.5	77.8	65.28	12.6	167.1
Luxembourg	2288.2	36.3	1.3	81.0	59.2	21.0	15.0	6.0	65.9

(continued)

Table 5.2 (continued)

	Government			Households		Other private entities			Non-domestic sources
	Direct payments to institution	Transfers to		Total expenditure	Of which	Total expenditure	Of which		Total expenditure
		Hcuseholds and students	Other private entities		Net payment to institution		Direct payments to institution	Transfers to households and students	
Hungary (7)	5445.0	260.4	0.0	1438.1	1204.7	43.0		43.0	27.6
Malta	583.7	35.6	26.3	117.6	36.6	9.4	3.4	0.0	4.6
Netherlands	36,007.2	5421.1	524.2	5534.8	3766.7	4494.3	4384.4	109.9	479.4
Austria	17,714.5	839.1	187.3	1577.0	555.6	167.5	157.5	0.0	0.0
Poland (8)	24.115.9	761.9	37.0	4259.1	3191.5	526.4	495.7	109.9	1 059.2
Portugal (8)	8929.2	481.5	0.0	2539.3	1875.9	91.6	91.6	0.0	436.8
Romania	6785.1	263.5	0.0	49.6	38.9	11.5	10.2	1.3	154.8
Slovania	2046.1	182.1	4.7	476.5	236.9	49.0	37.3	11.7	74.9
Slovakia (8)	3296.4	143.6	112.6	634.2	340.6	217.4	217.1	0.2	143.4
Finland	14,018.5	499.7	19.5	687.6	234.6	943.9	142.8	801.1	199.6
Sweden (9)	30,287.2	3373.5	0.0		473.4	797.6	795.4	0.0	342.7
Iceland (10)	1411.3	55.0	0.0	100.6	70.7	10.8	10.8	0.0	10.4
Norway	22,459.1	2858.5	321.2	709.4	709.4	468.5	147.3	321.2	97.5
Switzerland (11)	29,075.2	410.5	234.2				3212.2		177.3

(continued)

Table 5.2 (continued)

	Government			Households		Other private entities			Non-domestic sources
	Direct payments to institution	Transfers to		Total expenditure	Of which	Total expenditure	Of which		Total expenditure
		Households and students	Other private entities		Net payment to institution		Direct payments to institution	Transfers to households and students	
Serbia (12)	1408.2	0.0	0.0			0.0	0.0	0.0	0.0
Turkey	27,891.0	2275.8	0.0	8457.3	5696.9	4445.1	4271.4	173.6	181.7

(:) not available.
(1) Government and non- domestic *sources* definition differs.
(2) 2018.
(3) Definition differs.
(4) Excluding pro-primary education.
(5) Government, direct payments to institutions: 2016 and definition differs. Households: definition ciffers. Other private entities, total expendhure and transfors to households and students: 2017.
(6) Government and other private entities, transfers: 2016 and definition differs. Households: definition differs.
(7) Other private entities: 2018.
(8) Households, total expenditure definition differs.
(9) Other private emsties, tstal expendture and transfers: 2017.
(10) Government, direct payments to institutions: definition differs. Government transfers: 2017. Households, total expenditure: definition differs. Other private entities, transfers: 2018.
(11) Government and non-domestic *sources* 2018. Other private entities: 2016.
(12) 2017
Source Eurostat, https://ec.europa.eu/eurostat/statistics-explained/index.php?title=Educational_expenditure_statistics
Source Eurostat (online data codes: edoc_uoe_fine 01, educ_uoe_fine02 and educ_uoe_fine03)

having decreased due to the financial crisis. An increase in 2009 and, later, in 2020, up to 5.0% was to a large extent a result of the decrease in GDP due to the economic and financial crisis and to the COVID-19 pandemic.

At the level of the EU, the share of government expenditure on education out of total expenditure stood at 9.2% in 1995, increased to 10.4% at the beginning of the new century, then followed a decreasing trend: it decreased from 10.1% of total expenditure in 2019 to—as said above—9.4% of total expenditure in 2021, possibly due to the higher expenditures for other functions of the government.

Table 5.3 breaks down the total general government expenditure on education into the various levels of schools in 2021.

5.1.2 Cost Comparison

The amount of resources devoted by a society to education is one of the most important policy issues to be debated in a country, especially if one refers to public funds. A solution to it largely depends on whether these resources generate adequate returns, in either economic or non-economic terms, even if both are difficult to assess. An international comparison can be useful. This is offered by Fig. 5.4.

A couple of issues immediately rise when looking at these figures. The first refers to the factors that determine the differences in these values. The second is about whether these differences have an influence on the yield that can be derived by the various countries.

As to the former issue, deeper research on each specific country can reveal that in Luxembourg high teachers' salaries at the primary and secondary levels drive the costs up, whereas in Colombia these are very low, specifically due to the very high ratio of students to teaching staff (OECD, 2022).

In addition to this, differences in the expenditures as between the various countries depend on the existence of rules about continuing professional development for all teachers—and in some cases for heads of schools—as a regular part of their work, or for some teachers for specific purposes (e.g., promotion or salary increases) or on both. This is compulsory to some extent for teachers of general subjects at least at one level of education in most countries. The exception is offered by Denmark, Italy, the Netherlands and New Zealand.

As to the latter issue, to some extent econometric studies could say whether values such as the rate of growth of a country are related to the amount of expenses for education.[2] However, these exercises can be very difficult, due to the fact that some students in a country can be foreigners—as can happen in the case of Luxembourg, the United States and the United Kingdom—and the positive economic effects of

[2] The intuition behind this is based on the dictum expressed by the ancient Chinese philosopher and writer, Laozi, who said: "give a man a fish and he will eat for a day. Teach a man to fish and he will eat for a lifetime".

Table 5.3 Total general government expenditure on education in various levels of schools, 2021, % of GDP, *Source* Eurostat (2021)

	Education	Pre-primary and primary education	Secondary education	Post- Secondary and non-tertiary education	Tertiary education	Education not definable by level	subsidiary services to education	R&D Education	Education n.e.c
EU*	4.8	1.7	1.8	0.0	0.8	0.1	0.3	0.0	0.1
euro area 19*	4.7	1.5	1.9	0.0	0.7	0.1	0.3	0.0	0.1
euro area 20*	4.7	1.5	1.9	0.0	0.7	0.1	0.3	0.0	0.1
Belgium	6.3	2.0	2.4	0.0	0.9	0.6	0.2	0.0	0.1
Bulgaria	4.3	0.8	2.3		0.7	0.0	0.2	0.0	0.3
Czechia	5.1	1.3	2.3	0.0	0.7	0.1	0.2	0.3	0.1
Denmark	6.0	2.8	1.5	0.0	1.5	0.1	0.1	0.0	0.1
Germany*	4.5	1.5	1.6	0.1	0.8	0.1	0.4	0.0	0.1
Estonia	5.9	2.4	1.6	0.1	1.0	0.3	0.3	0.1	0.2
Ireland	3.0	1.2	1.1	0.0	0.6	0.0	0.1	0.0	0.0
Greece	4.1	1.3	1.2	0.0	1.0	0.0	0.1	0.4	0.1
Spain*	4.6	1.8	1.8	0.0	0.6	0.1	0.1	0.1	0.1
France*	5.2	1.4	2.2	0.0	0.6	0.2	0.7	0.0	0.0
Croatia	5.2	2.5	1.0	0.0	1.0	0.0	0.3	0.2	0.2
Italy	4.1	1.6	1.9	0.0	0.3	0.0	0.2	0.0	0.1
Cyprus	5.5	1.8	1.9	0.0	1.0	0.3	0.4	0.0	0.0
Latvia	5.6	2.2	1.3	0.0	1.0	0.5	0.1	0.1	0.4
Lithuania	4.8	1.0	1.9	0.2	0.6	0.4	0.0	0.2	0.4
Luxembourg	4.7	1.7	1.7	0.1	0.4	0.3	0.4	0.0	0.0
Hungary	5.0	1.2	1.5	0.0	1.6	0.1	0.4	0.0	0.2

(continued)

Table 5.3 (continued)

	Education	Pre-primary and primary education	Secondary education	Post- Secondary and non-tertiary education	Tertiary education	Education not definable by level	subsidiary services to education	R&D Education	Education n.e.c
Malta	5.5	1.4	1.9	0.0	0.9	0.0	0.4	0.4	0.5
Netherlands	5.1	1.6	2.0	0.0	1.3	0.0	0.2	0.0	0.0
Austria	4.9	1.5	2.0	0.0	0.8	0.3	0.2	0.0	0.1
Poland	4.9	2.3	0.9	0.0	1.2	0.1	0.3	0.1	0.1
Portugal*	4.6	1.7	1.8	0.0	0.7	0.1	0.1	0.0	0.2
Romania	3.2	0.8	1.3	0.0	0.7	0.1	0.1	0.0	0.2
Slovenia	5.7	2.3	1.8	0.0	1.1	0.1	0.2	0.0	0.1
Slovakia	4.3	1.2	1.5	0.0	0.6	0.3	0.5	0.0	0.1
Finland	5.7	1.3	2.4	0.0	1.7	0.2	0.0	0.0	0.1
Sweden	6.7	4.2	1.0	0.0	1.2	0.2	0.0	0.0	0.1
Iceland	7.7	3.4	2.3	0.0	1.5	0.1	0.2	0.0	0.1
Norway	5.0	2.2	1.0	0.0	1.1	0.4	0.2	0.0	0.1
Switzerland	5.7	1.3	1.6	0.0	1.3	1.2	0.1	0.0	0.1

Source Eurostat (2021)
Source Eurostat (gov_10a_exp)
* provisional

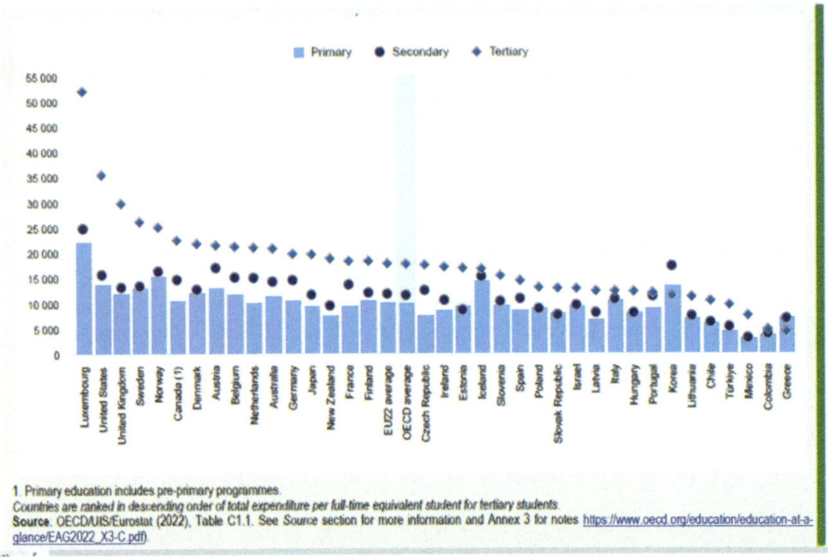

Fig. 5.4 Total expenditure per full-time equivalent student by level of education (2019) in equivalent US$ converted using PPPs for GDP; direct expenditure for educational institutions. *Source* OECD (2022)

their study can accrue to other countries or, by contrast, that positive yields can derive to a country from some of their young people studying abroad.

According to Goczek et al. (2021), educational skills—in particular the quality of primary and secondary education—are significant for GDP growth. Their analysis finds its source on various predecessors. One of them is offered by OECD analyses. In fact, according to OECD (2006), one additional year of education in the OECD area is estimated to have a long-term effect on the economic output between 3 and 6%. This can derive from rising labour productivity, which accounted for at least half of the per capita GDP growth in most OECD countries from 1994 to 2004. In addition, there can be a positive causal relationship between educational attainment and mental and physical health.

From Fig. 5.5 one can derive the relevance of education for growth of GDP per capita through the GDP per person employed.

In this figure changes in GDP per capita are broken down into demographic changes, labour utilisation and labour productivity. In most countries, improvements in the utilisation of available labour (i.e., increases in the share of the working age population that is employed) had a much larger impact on change in per capita output, but rising labour productivity accounted for at least half of per capita GDP growth in most OECD countries over the period 1994 to 2004. And this is due to a large extent to improvements in the quality of labour used in the production process, i.e., to the skills and competencies embodied in workers—or human capital—due to education.

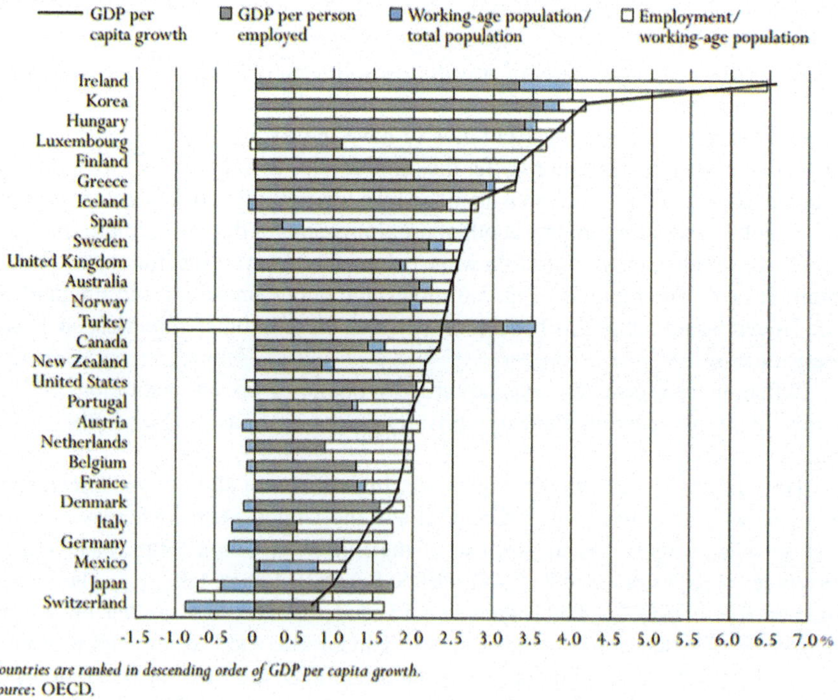

Fig. 5.5 The driving forces of GDP per capita growth (1994–2004). Trend series, average annual percentage change. *Source* OECD (2006)

However, it is not so easy to invest more in education. The expansion of educational opportunities and high-quality education implies higher current costs and possible reduction of other necessary expenses. To this end useful are international comparisons of expenditure for education and its determinants (whether for teachers' salaries at the various levels of education, teaching materials and facilities, number of students enrolled, class size, other services, particularly those related to ancillary services or R&D activities for the tertiary level, proportion of short-cycle tertiary programmes with respect to bachelor's, master's and doctoral levels) (OECD, 2020).

The *costs* of private and public educational institutions are similar for the OECD countries, with an average total expenditure for the primary to tertiary public institutions of about 11,900 $ per student, compared to under 12,100 $ for private institutions. Rather similar are differences in EU countries. However, there are substantial differences in some countries. In Israel and Turkey private expenditures are much higher than public ones, whereas the contrary happens in the Netherlands, the Czech Republic, Luxembourg, and New Zealand. Differences are even higher at the tertiary level, for which public institutions have expenditures more than 3 times higher than those of private institutions (14,100 $ against 4500 $ on average for the OECD).

The total average cost of education is 105,500 $ per student (OECD, 2022) (see Table 5.4).

The issue is that the costs of private institutions are charged to the families of the students, while those of public institutions are charged to taxpayers. Therefore, private institutions discourage families to have their sons attending their schools, assuming that these institutions offer the same service as public ones. Obviously, the fact that private schools are often attended by a lower number of students means that this is not the case, for problems both of proximity and for the educational content. In some cases, public institutions follow the behaviour of private institutions and charge high fees too. The following Fig. 5.6 shows that public universities in England and the United States adopt this behaviour, whereas those in other countries, such as in Italy, do not.

Different from costs are expenditures by public or private institutions. Private expenditures are spending funded by private sources, i.e. households and other private entities.

The number of students per class is very high in almost all Latin American countries, as well as in the United Kingdom and the United States, whereas it is very low in other countries such as those of the European Union (https://stats.OECD.org/Index.aspx?DataSetCode=EAG_PERS_RATIO). Let us take, e.g., the case of elementary schools, in which there are 15 students for every teacher on average across OECD countries, ranging from 9 to 1 in Greece and Luxembourg to over 23 to 1 in Brazil, Mexico, India (and, outside the OECD, the Russian Federation.) On average across OECD countries, the average class size differs between public and private schools by 1 student per class, both in primary and lower secondary education. On average, the number of students per class increases by 2 or more between primary and lower secondary education. The student-to-teacher ratio in lower and upper secondary education is slightly lower in private than in public institutions.

The following Fig. 5.7 shows that the expenditures for different levels of tertiary education in various OECD countries is very variegate. They vary from almost 60,000 $ in Luxembourg and around 30,000 $ in the U.K. to less than 5,000 $ in Greece for the upper level of tertiary education.

Expenses for primary education as a percentage of total educational expenditures are highest for Turkey (35%), Costa Rica (27%), Portugal (21%) and Mexico (15%), whereas those for lower secondary education are highest for Colombia (33%), Italy (32%), Spain (29%) and Mexico (27%). Considering the two lists of values, we can say that Mexico devotes the largest share of its expenses for education to primary and lower secondary education (15% for primary education and 27% for lower secondary education) (see Table 5.5).

As to the items of expenditures for education, core education services (such as teaching costs and other expenditures per student) represent 86% of the total from primary to tertiary educational institutions, whereas annual expenditure on R&D and ancillary services accounted for around 15% or more at primary to tertiary level.

Government transfers are used to offer incentives to institutions for better meeting student requirements, to increase access to education and reduce social inequalities.

Table 5.4 Public and total expenditure on educational institutions per full-time equivalent student, by type of institution (2019). *Source* OECD, 2022

Countries	Primary, secondary and Post-secondary non tertiary				Tertiary				Primary to tertiary			
	Public Sources		Total expenditure (public and private sources)		Public Sources		Total expenditure (public and private sources)		Public Sources		Total expenditure (public and private sources)	
	Public institutions	Private institutions	Public institutions	Private institutions	Public institutions	Private institutions	Public institutions	Private institutions	Public institutions	Private institutions	Public institutions	Private institutions
	(1)	(2)	(3)	(4)	(5)	(6)	(7)	(8)	(9)	(10)	(11)	(12)
OECD												
Australia	11,530	8105	12,150	13,099	8356	282	24,297	3383	10,685	7227	15,304	11,243
Austria	15,524	7783	15,787	11,920	21,483	9654	22,439	17,156	17,181	8647	17,021	14,330
Belgium	14,730	11,839	15,123	12,470	20,280	15,935	23,469	19,394	15,800	12,631	10,747	13,813
Canada[1]	11,302	2481	11,994	9569	11,990	a	22,335	0	11,481	2481	14,691	9580
Chile	6693	3755	6693	5585	9715	2891	15,436	9073	7143	3449	6139	5823
Colombia	3588	445	3593	4300	2923	0	7870	1216	3493	272	4204	3105
Costa Rioa[2]	5309	3032	m	m	m	m	m	m	m	m	m	m
Czeon Republic	10,011	3929	10,554	6931	14,804	584	18,919	4942	10,867	3083	12,047	6428
Denmark	12,169	9545	12,182	13,797	18,370	2778	21,617	33,308	13,783	9502	14 537	13,921
Estonia	8730	5923	8695	10,503	12,713	36	17,776	3241	9508	4461	10,531	8695
Finland	11,301	10,480	11,427	10,711	24,189	8756	28,927	9514	12,937	9511	13,331	10,035
France	m	m	m	m	m	m	m	m	m	m	m	m
Germany	m	m	m	m	m	m	m	m	m	m	m	m
Greece	6845	113	6840	9054	3130	2	4192	2	5463	113	5862	9654
Hungary	6805	6920	6957	12,248	8947	5132	12,467	9838	7079	6708	7965	11,958

(continued)

Table 5.4 (continued)

Countries	Primary, secondary and Post-secondary non tertiary				Tertiary				Primary to tertiary			
	Public Sources		Total expenditure (public and private sources)		Public Sources		Total expenditure (public and private sources)		Public Sources		Total expenditure (public and private sources)	
	Public institutions	Private institutions	Public institutions	Private institutions	Public institutions	Private institutions	Public institutions	Private institutions	Public institutions	Private institutions	Public institutions	Private institutions
	(1)	(2)	(3)	(4)	(5)	(6)	(7)	(8)	(9)	(10)	(11)	(12)
Iceland	14,021	10,143	15,048	11,038	10,249	9659	18,449	10,505	14,900	9942	15,832	10,813
Ireland	9038	3	9959	6400	11,931	2	10,830	23,859	9528	0	11,124	15,232
Israel	8066	9579	8253	13924	2996	7384	3107	14,583	7544	8513	8027	14,244
Italy	15,509	1011	10,892	6040	8522	1217	12,543	10,270	10,144	1459	11,211	7081
Japan	M	M	M	M	M	M	m	m	m	m	m	m
Korea	14,103	11,843	15,223	15,050	12,717	2232	12,465	9258	13,945	4784	15,711	10,794
Latvia	7013	4011	7293	10,930	7623	5952	8907	12,455	7028	8810	7331	12,303
Lithuania	6974	4462	7202	8867	8033	1568	11,535	6457	7208	3257	6150	7664
Luxemburg	24,917	8094	34,917	10,243	47,310	2	51,978	8	26,693	8094	27,083	16,243
Mexico	2693	17	2893	3956	5380	0	5911	8117	3100	11	3250	5414
Netherlands	11,340	3	11,704	81,492	15,904	2	20,023	23,211	12,325	8	13,670	30,288
New Zealand	7759	3778	8850	5205	10,764	3082	19,663	9358	6200	3667	10,793	5870
Norway	15,330	22,257	15,330	23,822	25,023	7894	27,101	14,320	17,368	15,991	17,593	19,677

(continued)

Table 5.4 (continued)

Countries	Primary, secondary and Post-secondary non tertiary				Tertiary				Primary to tertiary			
	Public Sources		Total expenditure (public and private sources)		Public Sources		Total expenditure (public and private sources)		Public Sources		Total expenditure (public and private sources)	
	Public institutions	Private institutions	Public institutions	Private institutions	Public institutions	Private institutions	Public institutions	Private institutions	Public institutions	Private institutions	Public institutions	Private institutions
	(1)	(2)	(3)	(4)	(5)	(6)	(7)	(8)	(9)	(10)	(11)	(12)
Poland	7810	5873	8572	10,294	13,085	2233	15,301	5977	8756	4222	9777	8473
Portugal	10,450	1322	10,760	5719	8731	109	12,117	10,734	10,101	1016	11,057	7731
Slovak republic	7153	5884	7733	7080	9750	559	13,290	7136	7578	5159	6640	7078
Slovenia	8963	5276	9916	7574	13,095	4813	18,274	6412	9747	5038	10,970	6977
Spain	10,169	4438	10,545	7550	11,745	912	14,934	11,677	10,588	3820	11,565	8298
Swedan	13,398	12,071	13,404	12,207	22,764	14,718	27,140	15,684	15,021	12,419	15,784	13,057
Switerland	10,350	15,771	M	M	30,065	9727	m	m	18,960	15,288	m	m
Turkiye	3914	223	4030	16,035	7716	1	8602	13,935	4572	347	4821	15,316
United kingdom	11,261	9921	11,932	13,015	0	7035	0	29,688	11,281	9122	11,932	17,316
United States	14,364	1539	14,902	12,411	15,392	5747	31,094	45,850	14,801	3552	17,945	28,413

(continued)

Table 5.4 (continued)

Countries	Primary, secondary and Post-secondary non tertiary				Tertiary				Primary to tertiary			
	Public Sources		Total expenditure (public and private sources)		Public Sources		Total expenditure (public and private sources)		Public Sources		Total expenditure (public and private sources)	
	Public institutions	Private institutions	Public institutions	Private institutions	Public institutions	Private institutions	Public institutions	Private institutions	Public institutions	Private institutions	Public institutions	Private institutions
	(1)	(2)	(3)	(4)	(5)	(6)	(7)	(8)	(9)	(10)	(11)	(12)
OECD average	10,316	6492	10,697	12,020	14,073	4549	17,955	13,453	11,304	5944	11,922	12,081
EUZZ average	10,690	5077	11,043	12,482	15,154	4743	18,463	13,017	11,385	5876	12,205	11,953
Partners												
Argentina	4192	1567	m	m	1(9)	1 (10)	m	m	5583	1798	m	m
Brazil	3873	3	m	m	15 385	2	m	m	4547	m	m	m
China	m	m	m	m	m	m	m	m	m	m	m	m
India	1554	450	m	m	m	m	m	m	m	m	m	m
indonesia	m	m	m	m	1811	105	m	m	m	m	m	m
Saudi Arabia	m	m	m	m	m	m	m	m	m	m	m	m
South Africa	2545	m	m	m	1M(9)	m	m	m	2945	m	m	m
G20 Average	7910	m	m	m	m	m	m	m	m	m	m	m

Source OECD, 2022

Note See Definitions and Methodology sections for more information. Data and more breakdowns are available at http://stats.oecd.org, Education at a Glance Database

1. Primary education includes pre-primary programmes. Post-secondary non-tertiary figures are treated as negligible
2. Year of reference 2020.
3. Year of reference 2018.

Source OECD/US/Eurostat. (2002). See Source section for more informtion and Annex 3 for notes (https://www.oecd.org/education\education-at-a-glance/EAG2022X3-C.Pdf). Please refer to the Reader's Guide for information concerning symbols for missing data and abbreviations.

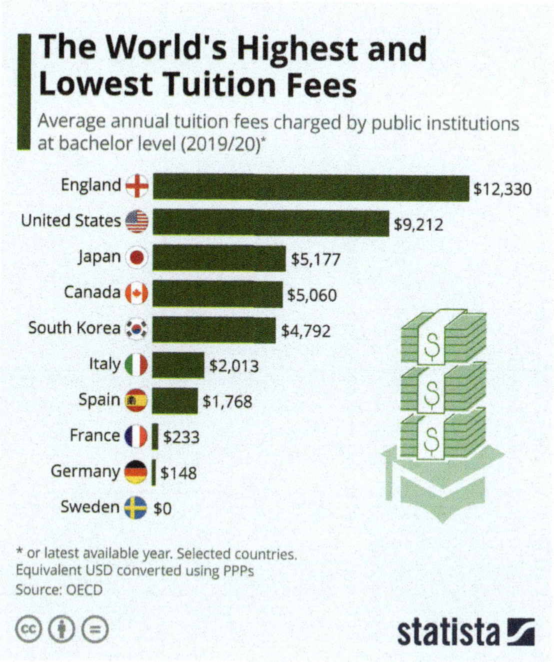

Fig. 5.6 World highest and lowest annual tuition fees charged by public institutions at the level of bachelor, 2019/2020. *Source* Statista, https://www.statista.com/chart/11058/bachelor-tuition-fees-international-comparison/

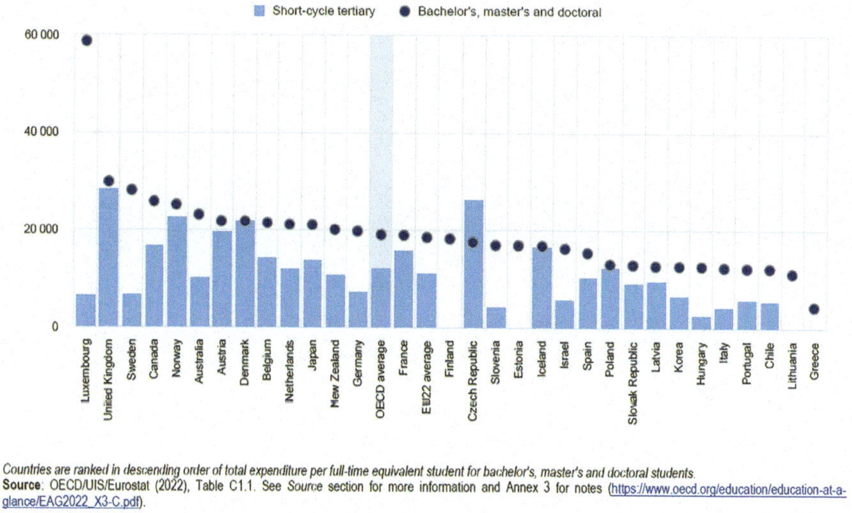

Fig. 5.7 Total expenditure per full-time equivalent student, by level of tertiary education (2019) in equivalent USD converted using PPPs; direct expenditure within educational institutions, various OECD countries. *Source* OECD (2022)

Table 5.5 Total expenditure on educational institutions per full-time equivalent student (2017) in equivalent US$, direct expenditure within educational institutions, by level of education. *Source* OECD, https://www.OECD-ilibrary.org/sites/5e4ecc25-en/index.hml?itemid=/content/component/5e4ecc25-en

		Below upper secondary					Upper secondary or post secondary non-tertiary		Tertiary				All levels of education
		Less than primary	primary	Completion of intermedate lower secondary programmes	Lower secondary	Completion of intermedate upper secondary programmes	Upper secondary	Post secondary non tertiary	Short Cycle tertiary	Bachelor's or equivalent	Master's or equivalent	Doctoral or equivalent	
		(1)	(2)	(3)	(4)	(5)	(6)	(7)	(8)	(9)	(10)	(11)	(12)
OECD	Countries												100
	Australia	0	4	a	12	a	29	6	11	28	9	2	100
	Austria	x(2)	1[a]	a	13	a	48	3	15	5	14	1	100
	Belgium	3	4	a	12	a	35	2	1	24	19	1	100
	Canada	x(2)	2[d]	a	5	a	21	10	26	24	11[c]	x(10)	100
	Chile	6	4	a	19	a	41	a	10	19	2[c]	x(10)	100
	Colombia	x(4)	x(4)	a	33[d]	5	31[d]	x(6)	x(9)	22[a]	x(9)	x(9)	100
	Coata Rica	11	27	8	8	3	18	0	7	15	3	0	100
	Czech Republic	0	0	a	5	a	68[d]	x(6)	0	7	19	1	100
	Denmark	x(2)	2[d]	a	16	a	40	0	5	21	15	1	100
	Estonia	0	0	a	10	a	39	9	6	14	21	1	100
	Finland	x(2)	2[d]	a	10	a	45	1	8	17	16	1	100
	France	2	3	a	13	a	41	0	14	12	14	1	100
	Germany	x(2)	4	a	10	a	41	13	1	18	11	2	100

(continued)

Table 5.5 (continued)

	Below upper secondary					Upper secondary or post secondary non-tertiary		Tertiary				All levels of education
	Less than primary	primary	Completion of intermediate lower secondary programmes	Lower secondary	Completion of intermediate upper secondary programmes	Upper secondary	Post secondary non tertiary	Short Cycle tertiary	Bachelor's or equivalent	Master's or equivalent	Doctoral or equivalent	
	(1)	(2)	(3)	(4)	(5)	(6)	(7)	(8)	(9)	(10)	(11)	(12)
Greece	1	10	a	9	2	34	9	1	25	8	1	100
Hungary	0	1	a	13	a	50	7	1	14	14	1	100
Iceland	x(2)	0[d]	a	21	a	29	7	4	22	16	1	100
Ireland	0	3	a	9	a	18	15	9	29	14	1	100
Israel	3	3	a	6	a	38	a	11	24	13	1	100
Italy	1	5	a	32	a	41	2	0	5	14	1	100
Japan	x(6)	x(6)	a	x(6)	a	44[d]	x(8)	21[d]	34[c]	x(9)	x(9)	100
Korea	x(2)	3[d]	a	6	a	39	a	14	33	4[d]	x(10)	100
Latvia	0	0	a	7	3	37	13	4	16	19	0	100
Lithuania	0	0	0	4	2	29	19	a	30	15	1	100
Luxembourg	2	6	a	11	a	28	2	4	15	29	2	100
Mexico	10	15	2	27	4	22	a	1	18	2	0	100

(continued)

Table 5.5 (continued)

	Below upper secondary					Upper secondary or post secondary non-tertiary		Tertiary				All levels of education
	Less than primary	primary	Completion of intermedate lower secondary programmes	Lower secondary	Completion of intermedate upper secondary programmes	Upper secondary	Post secondary non tertiary	Short Cycle tertiary	Bachelor's or equivalent	Master's or equivalent	Doctoral or equivalent	
	(1)	(2)	(3)	(4)	(5)	(6)	(7)	(8)	(9)	(10)	(11)	(12)
Netherland	2	4	a	13	a	37	0	2	24	16	1	100
New Zealand	x(4)	x(4)	a	19[d]	a	25	15	4	29	6	1	100
Norway	0	1	0	16	a	34	2	11	21	13	1	100
Poland	0	1	a	6	a	57	3	0	8	25	1	100
Portugal	1	21	a	19	a	27	1	0	9	21	1	100
Slovak Republic	0	1	0	6	0	63	2	0	4	23	1	100
Slovenia	0	0	a	8	a	51	a	8	9	18	5	100
Spain	2	5	a	29	a	23	0	12	11	16	1	100
Sweden	x(2)	3[d]	a	10	3	30	8	10	19	16	2	100
Switzerland	0	1	a	11	a	42[d]	x(6)	x(9,10,11)	24[d]	18[d]	3[d]	100
Turkiye	5	35	a	16	a	20	a	7	16	2	0	100

(continued)

Table 5.5 (continued)

		Below upper secondary					Upper secondary or post secondary non-tertiary		Tertiary				All levels of education
		Less than primary	primary	Completion of intermedate lower secondary programmes	Lower secondary	Completion of intermedate upper secondary programmes	Upper secondary	Post secondary non tertiary	Short Cycle tertiary	Bachelor's or equivalent	Master's or equivalent	Doctoral or equivalent	
		(1)	(2)	(3)	(4)	(5)	(6)	(7)	(8)	(9)	(10)	(11)	(12)
	United Kingdom	c	0	a	18	12	20	a	9	26	13	2	100
	United States	1	2	a	5	a	41[d]	x(6)	11	25	12	2	100
	OECD average	2	5	M	13	m	36	6	7	19	14	1	100
	EU22 average	1	3	M	12	m	40	6	5	15	17	1	100
Partners	Argentina	3	14	M	16	m	42	a	x(9)	23[d]	x(9)	1	100
	Brazil	11	17	a	14	a	38	a	x(9)	20[d]	1	0	100
	China	2	17	a	44	a	18	0	10	8	1[d]	x(10)	100
	India	35	12	a	30	a	8	1	x(9)	9[d]	x(9)	4	100
	Indonesia	11	27	a	19	a	30	a	3	5	5	0	100
	Saudi Arabia	m	m	m	m	m	m	m	m	m	m	m	m

(continued)

Table 5.5 (continued)

	Below upper secondary					Upper secondary or post secondary non-tertiary		Tertiary				All levels of education
	Less than primary	primary	Completion of intermedate lower secondary programmes	Lower secondary	Completion of intermedate upper secondary programmes	Upper secondary	Post secondary non tertiary	Short Cycle tertiary	Bachelor's or equivalent	Master's or equivalent	Doctoral or equivalent	
	(1)	(2)	(3)	(4)	(5)	(6)	(7)	(8)	(9)	(10)	(11)	(12)
South Africa	10	4	5	6	28	32	m	8	7	1d	x(10)	100
G20 average	m	10	m	16	m	30	m	11	17	9	1	100

Source OECD, https://www.OECD-ilibrary.org/sites/5e4ecc25-en/index.html?itemId=/content/component/5e4ecc25-en
Note Totals might not add up to 100% for the averages because of missing data for some levels for some countries. In most countries data refer to ISCED 2011. For Argentina and India data refer to ISCED-97. See Definitions and Methodology sections for more information. Data and more breakdowns are available at https:/istats.oecd.org/, Education at a Glance Database.
1. Year of reference differs from 2021: 2020 for Chile, China, India and South Africa.
Source OECD/ILO/UIS. (2022). See Source section for more information and Annex 3 for notes (https://www.oecal.org/education/education-at-a-glance/EAG2022_X3-A.pdf). Please refer to the Reader's Guide for information concerning symbols for missing data and abbreviations.

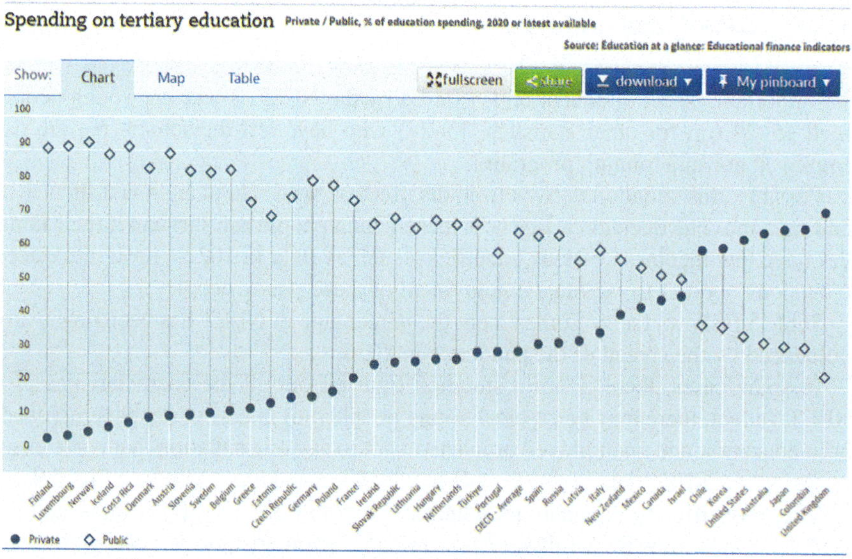

Fig. 5.8 Public and private spending for tertiary education, 2020, OECD countries. *Source* https://data.OECD.org/eduresource/spending-on-tertiary-education.htm

These transfers are not significant for non-tertiary educational levels, whereas in 2020 they represented almost 70% of the total funds devoted to tertiary education institutions, as an average at the level of the OECD, whereas they exceeded 80% of total expenditure in tertiary institutions of Northern and Central European countries (see Fig. 5.8).

5.2 Performance Metrics

In this section, we will deal with a number of topics, ranging from attendance to schools and NEET (young people *Not in education nor in employment or training*), proficiency and assessment methods, completion of studies, finding a job, social attainments and education, the role of teachers. A sub-section will be devoted to each of these topics, which represent aspects of the performance of the different educational set ups.

5.2.1 *Attendance and NEET*

An important aspect of education is offered by the NEET. These are young people aged 15–29 (or, for other statistics, 15–24) who have left the school, but are not employed nor in a training program.

Possibly, this situation derives from disaffection with school and the difficulty of finding a job. The question arises how these young people can survive, if their families have low incomes. Various countries—such as Italy in 2022—have introduced provisions in order to facilitate employment of these young people.

Table 5.6 shows that to some extent the percentage of NEET has been negatively influenced by the spread of the pandemic in 2020.

In some cases, there is not a situation figuring out all the characteristics of a NEET. In fact, there may be cases of young people: not in education, but employed; or in education and unemployed or inactive, not to say of a situation that is the exact opposite of a NEET, i.e., young people in education and employed (see Fig. 5.9).

There are different explanations of the existence and size of NEETs. Some authors, such as Assmann and Broschinski (2021), refer to various institutional configurations. In fact, care responsibilities are found to be an important factor in Central and Eastern European countries, where there are a lack of family-related services and weak formalised long-term care.[3] On the other hand, a low proportion of NEET having a disability prevails in Northern European countries, where generous and inefficient disability benefit schemes exist that can create false incentives to stay away from the labour market. By contrast, a high value of NEET can be found in countries having high labour market rigidities, low vocational specificity, and a lack of active labour market policies, such as Southern and some Central Eastern European countries. If this analysis is correct, country-specific measures must be taken to reduce the number of NEETs in Europe.

Caroleo et al. (2020) explain the propensity of an individual to be a NEET depending on his/her personal characteristics (gender, marital status, having children or not, education level, family background, condition of permanent inability and immigrant status), the educational and other institutions of the country of residence such as those of the labour market. By differentiating 19–24 and 25–30 year-old NEET, these authors have found that, while factors behind the former group were linked to the school-to-work transition process, those governing the latter were more closely linked to the functioning of the labour market. Finally, estimates referring to 2007 and 2016 allow to take account especially of the changes in the conditions of this market occurring before and after the financial crisis. 'In fact, the risk of being NEET, regardless of age and year of reference, is correlated positively with structural unfavourable conditions of the labour market and with passive subsidies to the unemployed, and negatively with the rate of growth of GDP, with active labour policies and with investments in education.' (Caroleo et al., 2020: 1020). From this point of view, labour policies such as job search assistance, short-term training courses and

---

[3] However, contrary to what one can imagine, the NEET rate is not higher for sons of very poor families. For this see Fig. 5.10 below.

Table 5.6 Trends of NEET, 2019–2021, OECD countries

		18–24 year-olds								
		2019			2020			2021		
		In education	Not in education		In education	Not in education		In education	Not in education	
			Employed	NEET		Employed	NEET		Employed	NEET
		(4)	(5)	(4) + (5) + (7)	(13)	(14)	(17) + (15) + (15)	(22)	(23)	(25) + (24) + (25)
OECD	Countries									
	Australia	m	m	m	m	m	m	m	m	m
	Austria	47	42	11.4	49	40	11.9	49	40	11.8
	Belgium	61	26	12.7	64	24	12.2	55	23	10.6
	Canada	39	47	13.5	40	41	19.7	41	44	14.9
	Chile	M	M	m	55	19	26.1	m	M	M
	Colombia	30	43	27.6	28	38	34.5	29	39	31.5
	Coata Rica	50	27	22.8	50	29	20.7	50	26	24.1
	Czech Republic	m	M	M	m	M	m	m	m	M
	Denmark	57	31	12.4	58	30	11.7	58	31	10.8
	Estonia	56	35	9.3	57	30	12.6	58	27	15.3
	Finland	55	32	12.8	55	31	14.8	58	30	11.9
	France	54	29	17.5	5	27	17.5	56	28	16.0
	Germany	62	29	8.2	62	28	9.4	61	29	9.7
	Greece	65	15	19.0	64	16	20.7	59	14	16.6

(continued)

Table 5.6 (continued)

Hungary	50	35	14.5	48	37	15.0	50	36	14.7
Iceland	56	37	6.6	60	31	9.0	59	33	7.7
Ireland	54	33	12.9	56	29	15.2	59	29	11.1
Israel	29	52	19.0	30	49	21.7	29	49	22.0
Italy	53	23	24.1	53	21	25.8	53	21	26.0
Japan	m	m	m	m	m	m	m	m	m
Korea	m	m	m	m	m	m	m	m	m
Latvia	61	28	10.2	60	29	10.7	59	28	13.2
Lithuania	61	28	11.6	60	25	15.0	53	32	15.1
Luxembourg	64	27	8.3	65	24	10.3	62	26	12.2
Mexico	38	41	21.5	38	39	23.3	39	39	21.8
Netherland	65	28	6.9	66	27	7.6	69	27	4.5
New Zealand	36	51	13.0	36	50	14.2	37	50	12.8
Norway	m	M	M	M	m	m	62	30	7.9
Poland	54	34	11.9	56	31	12.6	57	30	13.8
Portugal	54	33	12.9	58	27	14.5	62	25	13.5
Slovak Republic	56	32	12.5	57	29	13.8	59	27	14.3
Slovenia	63	27	9.6	66	23	10.6	71	20	8.4
Spain	58	23	19.2	59	19	22.0	61	20	19.2
Sweden	49	40	11.7	50	36	13.7	49	39	12.4
Switzerland	56	35	8.8	57	34	8.7	54*	34*	11.2*
Turkiye	35	31	33.9	35	28	36.9	37	31	32.5

(continued)

Table 5.6 (continued)

	United Kingdom	41	44	14.5	42	43	15.2	44	42	13.7
	United States	43	42	14.5	43	38	19.3	42	41	17.0
	OECD average	52	34	13.9	53	31	15.9	53	32	15.1
	EU22 average	57	30	12.9	58	28	14.2	59	28	13.4
Partners	Argentina	m	m	m	m	m	m	m	m	m
	Brazil	m	m	m	m	m	m	m	m	m
	China	m	m	m	m	m	m	m	m	m
	India	m	m	m	m	m	m	m	m	m
	Indonesia	m	m	m	m	m	m	m	m	m
	Saudi Arabia	m	m	m	m	m	m	m	m	m
	South Africa	41	15	43.8	43	12	45.0	m	m	m
	G20 average	m	m	m	m	m	m	m	m	m
		25–29 years olds								
		2019			2020			2021		
		In education	Not in education		In education	Not in education		In education	Not in education	
			Employed	NEET		Employed	NEET		Employed	NEET
		(31)	(32)	(35) + (33) + (34)	(40)	(41)	(44) + (42) + (43)	(49)	(50)	(53) + (51) + (52)

(continued)

Table 5.6 (continued)

OECD	Countries									
	Australia	M	M	M	M	M	M	M	M	M
	Austria	19	69	11.5	19	68	13.4	19	69	12.6
	Belgium	8	75	17.3	9	72	18.3	12	72	16.4
	Canada	10	75	14.3	10	70	19.8	11	73	15.9
	Chile	m	m	M	15	56	18.0	M	m	m
	Colombia	11	65	23.1	9	60	31.3	9	63	27.8
	Coata Rica	21	59	19.8	23	55	21.8	24	51	24.8
	Czech Republic	9	75	16.0	9	73	17.8	8	74	18.2
	Denmark	26	58	16.0	26	56	19.7	28	58	13.9
	Estonia	17	68	14.9	18	60	16.7	17	69	13.5
	Finland	28	59	13.4	28	59	13.5	29	57	14.0
	France	8	71	20.6	8	70	21.6	10	71	19.3
	Germany	21	68	11.7	21	65	13.0	21	65	12.9
	Greece	15	57	28.0	17	53	29.6	17	53	29.5
	Hungary	7	75	17.1	8	72	19.2	7	79	13.7
	Iceland	26	63	9.0	28	61	11.2	26	62	12.6
	Ireland	13	72	15.3	12	68	19.7	16	68	15.4
	Israel	29	57	14.0	27	54	18.9	27	54	18.9
	Italy	16	53	30.7	17	50	33.3	16	53	31.2

(continued)

Table 5.6 (continued)

Japan	m	m	m	m	m	m	m	m	m
Korea	m	m	m	m	m	m	m	m	m
Latvia	10	76	14.6	9	70	20.4	10	70	19.9
Lithuania	8	77	15.3	7	75	18.0	7	78	15.7
Luxembourg	16	73	11.0	14	74	11.9	16	74	10.8
Mexico	8	65	25.1	8	64	28.2	8	65	27.3
Netherland	22	69	9.4	22	69	9.2	23	70	7.7
New Zealand	12	74	13.6	13	73	14.6	14	73	12.9
Norway	m	m	m	m	m	m	23	67	10.2
Poland	6	76	18.0	5	75	19.7	5	78	17.2
Portugal	12	75	13.1	11	72	16.5	15	69	16.3
Slovak Republic	6	73	20.8	6	71	22.1	7	73	19.4
Slovenia	18	70	12.8	17	70	12.6	m	m	m
Spain	15	60	24.4	16	56	28.2	16	58	25.8
Sweden	23	67	9.6	24	65	10.5	25	65	9.8
Switzerland	22	71	7.9	22	70	7.3	21*	68*	10.4*
Turkiye	15	50	35.2	15	47	38.2	15	50	35.0

(continued)

Table 5.6 (continued)

	United Kingdom	10	77	12.9	11	76	12.9	12	75	12.6
	United States	11	73	16.2	11	67	21.7	11	70	19.0
	OECD average	15	69	16.2	15	65	18.7	16	56	17.6
	EU22 average	15	69	16.4	15	67	18.3	15	58	16.8
Partners	Argentina	m	m	m	m	m	m	m	m	m
	Brazil	m	m	m	m	m	m	m	m	m
	China	m	m	m	m	m	m	m	m	m
	India	m	m	m	m	m	m	m	m	m
	Indonesia	m	m	m	m	m	m	m	m	m
	Saudi Arabia	m	m	m	m	m	m	m	m	m
	South Africa	7	40	52.2	7	35	57.4	m	m	m
	G20 average	m	m	m	m	m	m	m	m	m

Note NEET refers to young People who are neither employed nor in formal education or training. See Definitions and Methodology sections for more information. Data and more breakdowns are available at http://stats.Orcd.org/, Education at a Glance Database

Source OECD (2022)

Additional columns showing data on the percentage of young adults in education by work status, and the percentage of unemployed and inactive NEETs are available for consultation on line (see StatLink below).

Source OECD (2022). See Source section for more information and Annex 3 for notes (https://www.oced.org/education/education.at.a.glance/EAG.2022X3-A.pdf)

Please refer to the Reader's Guide for information concerning symbols for missing data and abbreviations

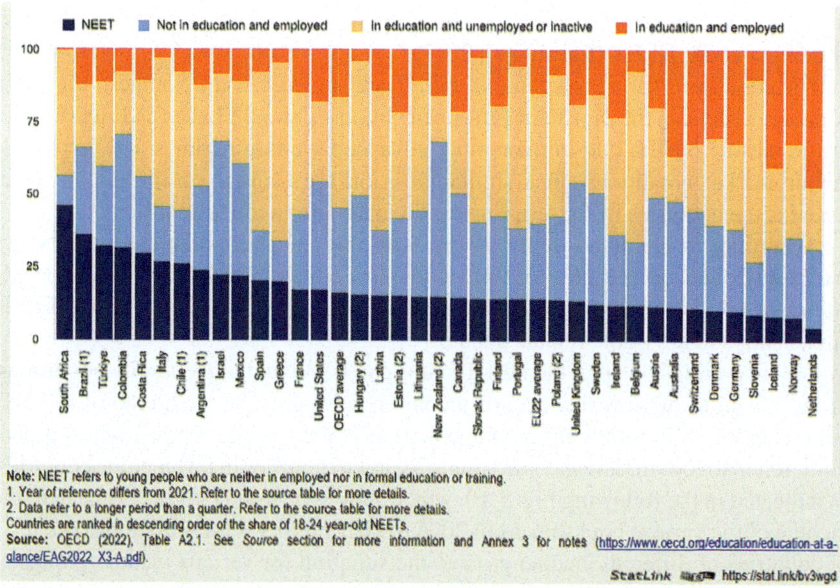

Fig. 5.9 Distribution of 18–24 years-old by education and work status (2021, quarterly data), %. *Source* OECD (2022)

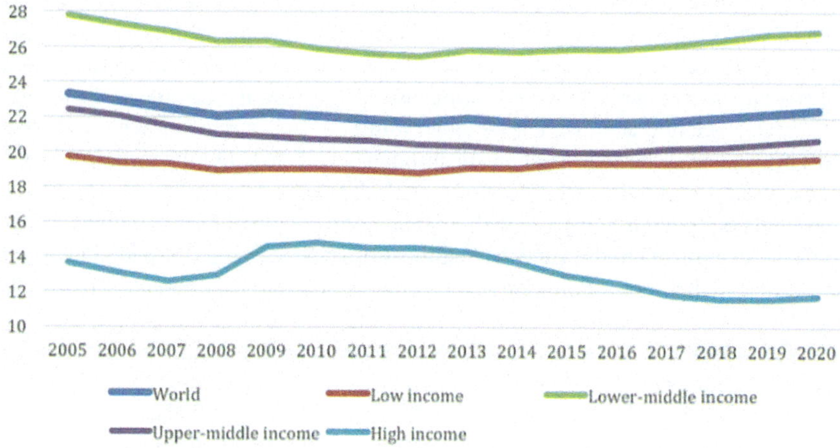

Fig. 5.10 Youth NEET rates globally and by income groups of countries, 2005–2020. *Source* ILO (2019). *Note* In descending order, the curves indicate: lower-middle income countries, the world, upper-middle income countries, low income countries and high-income countries

subsidised work together with other active labour market policies can lower the risks of being a youth NEET (Eurofound, 2012).

The conditions for a high value of NEET are satisfied in the Southern European countries, whereas the Northern European countries should have (and really have) low values of NEET, due to their more advanced welfare regimes and favourable labour market conditions. In a similar position are Central continental.countries whereas the countries of Eastern Europe lie in an intermediate position.[4]

According to Quintano et al. (2018), who refer to the Italian case, the NEET are mainly older people, have parents with a low educational level or young people attending specific fields of study, reside in the South of Italy, don't have Italian citizenship and are of the female gender.

We have said above that the NEET status can be related to the family conditions. From this point of view the level of income is relevant. The likelihood of a NEET belonging to families in high-income countries is the lowest, but the highest probability refers to low-middle income countries, not to families in low income countries, as indicated in the following Fig. 5.10, which shows the trend of NEET related to the income of the various countries from 2005 to 2020. If this is the situation at the level of countries of different income groups, the situation for various income groups of families in a given country might be different. In fact, it may well be that low-income families have a higher NEET rate than lower-middle families.

From a research referred to Armenia and Georgia it results that 'youth in wealthier households have a higher likelihood of being non-NEET because they can afford to attend schools and/or have better networks in the labor market' (Buitrago-Hernandez et al., 2019). In addition, similar results derive from a research referred to quite a different environment, that of Florida. Here low 'family income is associated with poor academic achievement among children. Higher rates of school absence and tardiness may be one mechanism through which low family income impacts children's academic success' (Morrissey et al., 2014).

A final notation must be given with reference to a couple of factors influencing the level of education of children and the qualification of NEETs, i.e., the parental level of education and the status of migrant of the young people and their parents. Figure 5.11 shows, on the left-hand side, that the NEETs that are sons of parents with medium- or high-levels of education are very few (respectively, 6.9% and 2.9%), whereas the number of young people whose parents have a low level of education is much higher (26.1%), i.e., around 9 times more likely to be early school leavers than young people having parents with a high level of education. As to the status of migrants of young people and their parents, if these were born in another EU country or outside the EU, their probability of being a NEET is 23.8% or 22.6%, respectively, whereas the probability is much lower if the young people were born in the country and/or their parents in another EU country or—even more—in the reporting country. The Fig. 5.11 also shows that.

The NEET status is not a temporarily problematic situation and there are both microeconomic and macroeconomic factors behind the propensity of an individual to

[4] Similar are the results obtained by Brunello, De Paola (2014).

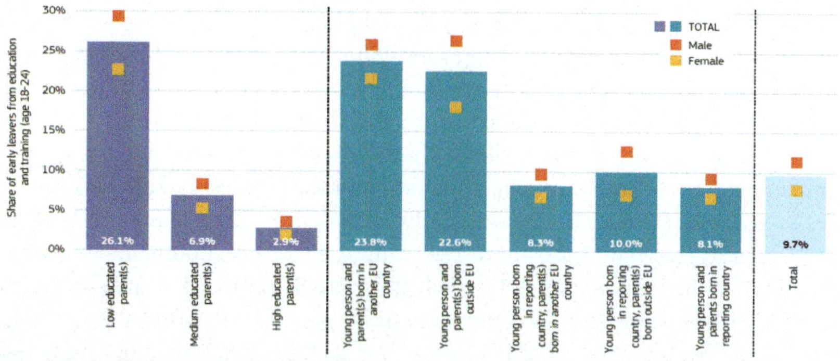

Source: Eurostat (EU Labour Force Survey 2021 ad hoc module for parental education and EU Labour Force Survey 2021 for parental country of birth and total).
Download data Monitor Toolbox
Note: parental education denotes the highest level of education successfully completed between the father and the mother of the respondent; low reliability for females born in the reporting country with parent(s) born outside the EU.

Fig. 5.11 New evidence sheds light on parental education and parental country of birth of NEET. *Source* European Commission (2022)

be a NEET. The former are the personal characteristics of the young people and refer to the factors already indicated (following Caroleo et al. 2020). Thus, the transition from school to work is rather easy for the highly educated young people, for whom the NEET status is often of short duration.[5] The latter—i.e., macroeconomic factors - refer to labour market features, such as the current rate of unemployment and the rate of growth of GDP as well as existence of passive subsidies for the unemployed. The latter condition implies that the Northern European countries—followed by the countries of Central Europe—are those where the rate of NEET is lower. At the same time, due to the same factors, Mediterranean countries, such as Italy,[6] have a high rate of NEET, due, first, to high rates of unemployment and also, to some extent, to the difficulties of teaching in an efficient way, deriving—among other factors—from the scarcity of teachers and the low background of some of them, due also to their low pay.

For these reasons one can say that the presence of high values of NEET indicates the existence of problems, at the level not only of individuals, but also of the country where they live. A low value of NEET can be desired for equity and efficiency considerations. From the former point of view, reducing this value is desirable, since NEET rates are higher for the disadvantaged section of the population. From the latter point of view, one can say that early leaving of the school, on the one side, reduces the amount of human capital available in the country, which could offer an important contribution to production and, on the other hand, the level of production is reduced, due to the reduction of the positive externalities that could be created by education. A higher education is also important for the economic and social system, as it reduces

[5] It is also worth recalling that the NEET rate is higher for young people having lower secondary education at most (EU Commission, 2022).

[6] There are 3 such countries (Italy, Greece and Spain) among the first 6 European countries as to the percentage of NEETs.

the probability of individuals engaging in socially costly activities, such as crime[7] (Brunello and De Paola, 2014). Therefore, the role of the school in enriching the education offer and learning opportunities is crucial as a focal point and learning hub and should therefore be strengthened (European Commission, 2022).

Analysis can distinguish the role of malfunctioning in the education system as separate from that of the improper functioning of the labour market. This has been done, in particular with reference to Italy and Spain. These countries have high rates of NEET, due mainly to these two causes, with a predominant role played, respectively in the two countries, by the former and the latter, as shown by De Luca et al. (2020). Investing more in education and adopting measures able to educate young people for a longer period is a must for many countries having high values of NEET. Also clarification of the economic advantages of more education from the personal point of view is needed. Caroleo and Pastore (2019) have shown that the probability of being employed in Italy is much higher for graduates (76.2%) with respect to that of people having only a secondary school diploma (30.1%). Thus, higher educational attainment leads to better labour-market outcomes. This is true—even if to a less extent—also for other OECD countries. In fact, in 2021, on average, OECD employment rates for tertiary-educated people aged 25–34 were 8 p. p. higher than those who had attended only upper secondary schools and 26 p. p. higher than those with lower than upper secondary education[8]; at the same time, higher educational attainment is negatively correlated with rates of unemployment and labour-market inactivity. Even if tertiary attainment among 25–34 years-old has increased, its labour-market value is practically identical to the one in 2000.

The highest average employment rates across OECD countries are for tertiary-educated individuals having studied information and communication technologies, and the lowest are for those who studied arts and humanities, social sciences, journalism and information, even if also graduates in these subjects have higher employment rates than people having attained only an upper secondary or post-secondary title.

Due to the nature of the determinants of high rates of NEETs, the consequences of these high rates are long-lasting. They are of a psychological, material and behavioural nature. In fact, the development of people's identity can be negatively affected from a psychological perspective. Psychological distress, such as feelings of loneliness, powerlessness, restlessness, anxiety and depression are common in NEETs. These are usually less skilled individuals who, being excluded from production and education, are more likely to experience in the future less continuous and

[7] In fact, more in general NEETs can be involved in risky behaviour, including also alcohol and drug abuse, leading to a dangerous spiral.

[8] In 2021, the average unemployment rate of tertiary-educated younger adults was 2 p. p. below that of their peers with upper secondary or post-secondary non-tertiary attainment and 9 p. p. lower than for those with below upper secondary attainment.

lower paid employment than other people, poor employment outcomes and life-time earnings.[9]

With reference to the proper policy strategies to reduce their number, one can suggest: an appropriate education system, bridging courses or second chance education programs, existence of school-to-work transition institutions—of the kind of apprenticeship—and temporary working contracts for the NEETs (European Commission, 2022). To some extent similar suggestions come from Brown et al. (2022). Finally, from a long-term perspective (without which all these mentioned measures have no significant effect), only an increase in investment is effective, as it can raise productivity, growth and labour demand.

5.2.2 *Proficiency and Assessment Methods*

As said, PISA measures 15-years-old's ability to use students' reading, mathematics and science knowledge and skills to meet real-life challenges. The results of the 2022 tests are shown by Table 5.7. In previous PISA tests the top-ranking countries were the same as in 2018, i.e. China, Singapore, Macau, Hong Kong, Taiwan and Japan.

In some countries, in addition to the assessments provided by PISA tests there are other enquires to evaluate the educational system of these countries. We will refer only to a few countries, Finland, Italy, the United Kingdom and the United States.

In the case of Finland, an institution, the FINEEC, exists whose main purpose is to assess and evaluate learning outcomes, in pre-tertiary schools and also in higher education institutions in order to develop education and to support its quality (Mattarelli and De Santis, 2020).

In Italy tests are provided by INVALSI, which is the Italian National Institute for the Evaluation of the Educational and Training System. This carries on annual checks of the results of the learning process of Italian students. It also devises instruments for self-assessment of schools. Students are submitted to three different tests: Italian, Mathematics and English (Pileggi, 2018).

In the United Kingdom there are *Standard Assessment Tests* (SATs) since 1926, or national curriculum assessments, referring to the attainment and progress of primary pupils, in various subjects. At the age of 10–11, these cover English, mathematics, and science. The results of these assessments are published and are thus available also to families. Before that age other assessments are performed. The United Kingdom Government plans to pursue a level of 90% for reaching the expected standards in those subjects by 2030 (Williams, 2023).

Also in the United States there is since 1969 a *National Assessment of Educational Progress* (NAEP), which is the largest assessment of knowledge for American

[9] As to the costs deriving from the existence of NEET in terms – among the various items - of social exclusion, crime and educational underachievement, also of their children, one can see: Coles et al. (2002, 2010), Prince's Trust (2010), Helliwell and Huang (2011).

Table 5.7 PISA 2022 Results for mathematics, science and reading, 2018. OECD (2023), PISA 2022 Results: Combined Executive Summaries, Volumes 1, 2 & 3

	Mathematics	Reading	Science	Mathematics	Reading	Science	Mathematics	Reading	Science	One subject (Level 5 or 6}	Three subjects (below Level 2)
	Mean	Mean	Mean	Score diff.	Score diff.	Score diff.	Score diff.	Score diff.	Score diff.	%	%
OECD Average	472	476	485	–7	–4	–7	–15	–10	–2	13.7	16.4
Singapore	575	543	561	6	12	12	6	–7	10	44.5	4 2
Japan	536	516	547	2	2	4	9	12	17	28.7	5.3
Korea	527	515	528	–13	–11	–4	1	1	9	29.7	7.3
Estonia	510	511	526	1	11	–3	–13	–12	–4	20.0	5.2
Switzerland	508	483	503	–12	–7	–11	–7	–1	7	19.4	12.4
Canada*	497	507	515	–17	–9	–12	–15	–13	–3	22.7	8.1
Netherlands*	493	459	488	–20	–25	–23	–27	–26	–15	19.0	20.2
Ireland*	492	516	504	–2	–1	–7	–8	–2	8	14.7	7.5
Belgium	489	479	491	–18	–11	–11	–19	–14	–8	15.5	15.2
Denmark*	489	489	494	–9	0	–3	–20	–12	1	12.8	10.3
United Kingdom*	489	494	500	–1	2	–10	–13	–10	–5	17.9	12.0
Poland	489	489	499	5	5	–1	–27	–23	–12	15.3	11.9
Austria	487	480	491	–9	–5	–14	–12	–4	1	14.6	15.5
Australia*	487	498	507	–21	–14	–16	–4	–5	4	20.7	12.1

(continued)

Table 5.7 (continued)

	Mathematics	Reading	Science	Mathematics	Reading	Science	Mathematics	Reading	Science	One subject (Level 5 or 6}	Three subjects (below Level 2)
	Mean	Mean	Mean	Score diff.	Score diff.	Score diff.	Score diff.	Score diff.	Score diff.	%	%
Czech Republic	487	489	498	−12	1	−9	−12	−2	1	15.5	12.2
Slovenia	485	469	500	−7	−7	−10	−24	−27	−7	13.0	12.0
Finland	484	490	511	−34	−23	−34	−23	−30	−11	17.9	11.5
Latvia*	483	475	494	2	3	−1	−13	−4	7	9.7	10.6
Sweden	482	487	494	−9	−11	−2	−21	−19	−6	17.0	15.2
New Zealand'	479	501	504	−24	−12	−18	−15	−5	−4	19.5	13.7
Lithuania	475	472	484	−4	2	−6	−6	−4	2	10.4	14.4
Germany	475	480	492	−12	2	−17	−25	−18	−11	14.6	16.7
France	474	474	487	−14	−8	−6	−21	−19	−6	12.9	16.8
Spain	473	474	485	−4	−1	−2	m	m	m	10.6	12.9
Hungary	473	473	486	−10	−5	−15	−8	−3	5	11.2	16.5
Portugal	472	477	484	8	7	5	−21	−15	−7	10.1	13.8
Italy	471	482	477	8	1	−6	−15	5	9	10.7	12.9
Viet Nam**	469	462	472	m	m	m	m	m	m	6.3	12.2
Norway	468	477	478	−7	−5	−7	−33	−23	−12	13.8	17.5

(continued)

Table 5.7 (continued)

	Mathematics	Reading	Science	Mathematics	Reading	Science	Mathematics	Reading	Science	One subject (Level 5 or 6}	Three subjects (below Level 2)
	Mean	Mean	Mean	Score diff.	Score diff.	Score diff.	Score diff.	Score diff.	Score diff.	%	%
Malta	466	445	466	3	3	2	–6	–3	9	10.7	21.6
United States*	465	504	499	–8	2	5	–13	–1	–3	18.1	14.8
Slovak Republic	464	447	462	–16	–13	–20	–22	–11	–2	9.5	22.2
Croatia	463	475	483	–1	0	–10	–1	–3	10	9.7	13.6
Iceland	459	436	447	–24	–24	–27	–36	–38	–28	6.8	23.3
Israel	458	474	465	11	13	7	–5	3	3	15.1	21.3
Turkiye	453	456	476	14	5	24	0	–10	8	7.3	18.5
Brunei Darussalam	442	429	446	m	m	m	12	21	15	4.5	30.0
Serbia	440	440	447	3	16	4	–8	1	8	5.0	24.5
UnitedArab Emirates	431	417	432	7	–12	–8	–4	–14	–2	8.8	33.9
Greece	430	438	441	–9	–12	–21	–21	–19	–11	3.9	25.7
Romania	428	428	428	6	15	3	–2	1	2	5.0	33.2
Kazakhstan	425	386	423	10	–4	6	2	–1	26	2.2	32.8
Mongolia	425	378	412	m	m	m	m	m	m	2.3	39.9

(continued)

Table 5.7 (continued)

	Mean score in PISA 2022			Long term trend. Average decimal trend			Short term change in performance (PISA 2018 to PISA 2022)			Top-performing and low-performing students	
	Mathematics	Reading	Science	Mathematics	Reading E-T777T"H	Science	Mathematics	Reading	Science	Share of top performers in at least one subject (Level 5 or 6)	Share of low performers in all three subjects (below Level 2)
Bulgaria	417	404	421	3	–5	–11	–19	–16	–3	4.6	38.3
Moldova	414	411	417	14	20	5	–6	–13	–12	1.7	37.1
Qatar	414	419	432	58	59	51	0	12	13	5.2	34.2
Chile	412	448	444	–1	16	2	–6	–4	0	3.6	24.8
Uruguay	409	430	435	–8	3	5	–9	3	10	3.4	30.6
Malaysia	409	388	416	7	–12	1	–32	–27	–21	1.3	40.6
Montenegro	406	405	403	10	9	0	–24	–16	–12	1.5	41.3
Mexico	395	415	410	2	4	1	–14	–5	–9	0.7	38.4
Thailand	394	379	409	–8	–20	–8	–25	–14	–17	1.3	46.3

(continued)

Table 5.7 (continued)

	Mean score in PISA 2022			Long term trend. Average decimal trend			Short term change in performance (PISA 2018 to PISA 2022)			Top-performing and low-performing students	
	Mathematics	Reading	Science	Mathematics	Reading E-T777T"H	Science	Mathematics	Reading	Science	Share of top performers in at least one subject (Level 5 or 6)	Share of low performers in all three subjects (below Level 2)
Peru	391	408	408	26	38	33	–9	8	4	1.3	40.8
Georgia	390	374	384	8	–2	6	–8	–6	1	1.3	51.1
SaudiArabia	389	383	390	m	m	m	16	–17	4	0.3	48.6
North Macedonia	389	359	380	m	–2	m	–6	–34	–33	0.7	55.8
Costa Rica	385	415	411	–17	–21	–16	–18	–11	–5	1.1	38.1
Colombia	383	409	411	9	12	15	–8	–4	–2	1.5	40.7
Brazil	379	410	403	10	7	5	–5	–3	–1	2.6	42.2
Argentina	378	401	406	–5	–2	7	–2	–1	2	1.5	42.7
Jamaica*	377	410	403	m	m	m	m	m	m	1.7	43.5
Albania	368	358	376	4	12	–5	–69	–47	–41	0.8	56.2
Indonesia	366	359	383	0	–5	0	–13	–12	–13	0.1	59.0
Morocco	365	339	365	m	m	m	–3	–20	–11	0.0	68.5
Uzbekistan	364	336	355	m	m	m	m	m	m	0.1	71.4
Jordan	361	342	375	–8	m	m	–39	m	m	0.0	62.9
Panama*	357	392	388	–4	15	5	4	15	23	1.2	50.4

(continued)

Table 5.7 (continued)

	Mean score in PISA 2022			Long term trend. Average decimal trend			Short term change in performance (PISA 2018 to PISA 2022)			Top-performing and low-performing students	
	Mathematics	Reading	Science	Mathematics	Reading E-T777T"H	Science	Mathematics	Reading	Science	Share of top performers in at least one subject (Level 5 or 6)	Share of low performers in all three subjects (below Level 2)
Philippines	355	347	356	m	m	m	2	7	–1	0.2	71.3
Guatemala	344	374	373	m	m	m	10	5	8	0.1	63.8
El Salvador	343	365	373	m	m	m	m	m	m	0.2	62.8
Dominican Republic	339	351	360	m	m	m	14	10	25	0.1	68.4
Paraguay	338	373	368	m	m	m	11	3	10	0.1	61.1
Cambodia	336	329	347	m	m	m	12	8	17	0.0	82.2
Macao (China)	552	510	543	18	14	24	–6	–15	0	31.1	4.1
Chinese Taipei	547	515	537	–6	8	2	16	13	22	34.8	7.9
Hong Kong (China)*	540	500	520	–3	–5	–21	–11	–25	4	29.7	7.2
Ukrainian regions (18 of 27)	441	428	450	m	m	m	m	m	m	4.6	25.3
Cyprus	418	381	411	m	m	m	–32	–43	–28	5.3	40.3

(continued)

Table 5.7 (continued)

	Mean score in PISA 2022			Long term trend. Average decimal trend			Short term change in performance (PISA 2018 to PISA 2022)			Top-performing and low-performing students	
	Mathematics	Reading	Science	Mathematics	Reading E-T777T"H	Science	Mathematics	Reading	Science	Share of top performers in at least one subject (Level 5 or 6)	Share of low performers in all three subjects (below Level 2)
Baku (Azerbaijan)	397	365	380	m	m	m	–23	–24	–18	0.9	50.9
Palestinian Authority	366	349	369	m	m	m	m	m	m	0.1	63.5
Kosovo	355	342	357	m	m	m	–11	–11	–8	0.0	72.9

Source OECD (2023), PISA 2022 Results: Combined Executive Summaries, Volumes I, II, & III
Note Medium general orientation is not available in Luxembourg due to very small sample size
Source Eurostat (online data code: edat_lfe_9903)
Notes Values that are statistically significant are marked in bold (see Annex A3). * Caution is required when interferring estimates because one or more PISA sampling standards were not

students in various subjects, useful for providing information to educators, policymakers, elected officials, and parents about how children learn in each school, also in comparison to other children elsewhere in the States. In 2022/2023, the NAEP schedule has included assessments of Mathematics and Reading at the age of 13 in Fall 2022, and a field test at grades 4, 8, and 12 in Spring 2023. Other tests, less frequent, have as their object science, writing, arts, civics, economics, geography, technology and engineering literacy, and U.S. history The NAEP policy is set by the *National Assessment Governing Board* (NAGB), which is an independent, bipartisan board, whose members are appointed by the U.S. Secretary of Education.[10] In addition to assessing student achievement in various subjects, NAEP also surveys students, teachers, and school administrators to help provide contextual information. NAEP results are useful as teachers, principals, parents, policymakers, and researchers can use them to assess student progress and develop ways to improve education (Mullis, 2019).

5.2.3 Completion of Studies

Figure 5.12 shows the distribution of the population aged 25–74 years in 2022 in the European Union, according to the educational level attained: less than primary, primary or lower secondary level of education (indeed, a low educational attainment level); upper secondary or post-secondary non-tertiary education (a medium level of education, both for general or vocational orientation); tertiary education (i.e., a high level of education). It says that the EU countries as a whole had a percentage of low-level educated people lower than that of Greece, Italy, Malta, Portugal and Spain. Most Nordic countries, Slovenia, Austria, Ireland, Cyprus, Luxembourg, France, Belgium, the Netherlands and Spain had, instead, a level of education above the average.

Figure 5.13 shows the trend of this population in the last decade. The most important indication is that the fraction of people with a tertiary education has risen, to the expenses of both the other levels.

To show the tertiary education level of the new generations, it may be useful to look at Fig. 5.14.

This figure shows that the tertiary education attainment level is higher or lower than both the 2022 average and the target set for 2030—which is a bit higher—in countries belonging to various groups. To make some examples: Finland, which is part of the Scandinavian system, is below both averages, in contrast with all the other countries of this system. Spain and France are above the averages, in contrast to Italy. Greece is slightly above the 2022 EU average, even if a little below the 2030 target.

Table 5.8 below can be useful to indicate the levels of education for various ages of the population, i.e., those aged 25–54 and 55–74. This table confirms that higher

[10] Its members include representatives of: government and public policy (of different political parties); teachers and principals, experts; the general public (again, of different political parties).

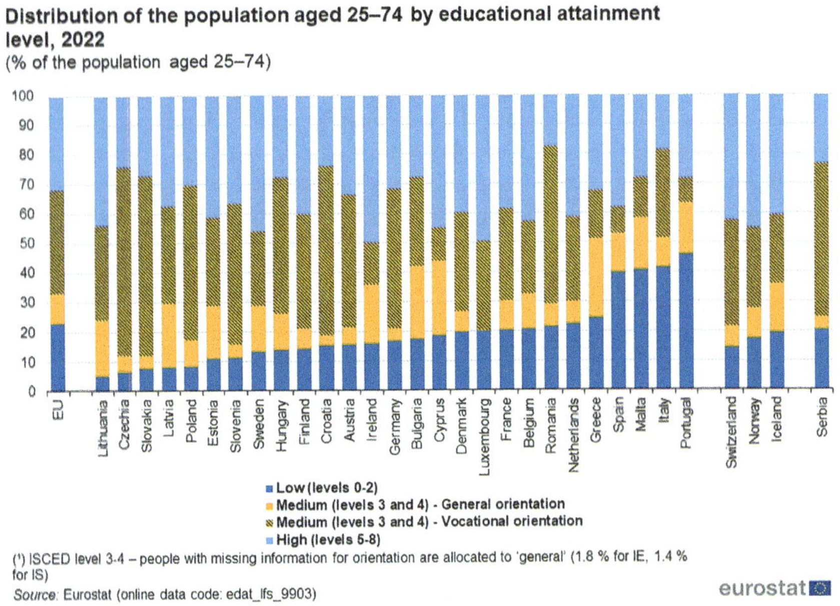

Fig. 5.12 Education level of population aged 25–74, various European countries, 2022. *Source* Eurostat (2023). *Note* The various levels of education for each country and the EU are indicated by the various histograms starting from the low up to the high level when going from the bottom to the top

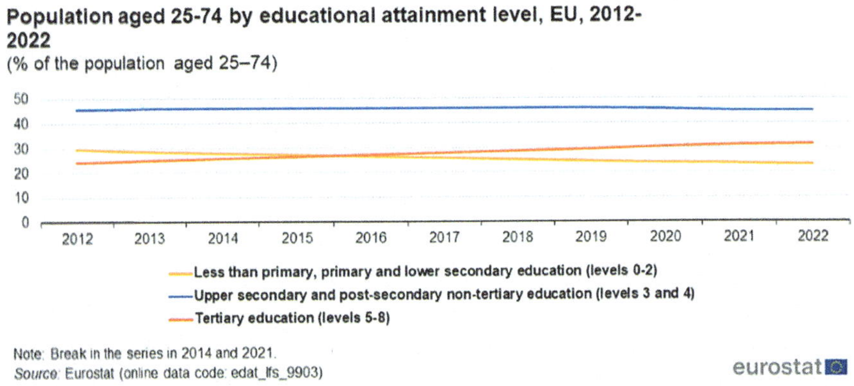

Fig. 5.13 Trend of the population aged 25–74 with the various educational attainment levels, 2012–2022, %. *Source* Eurostat (2023). *Note* The curve always above indicates the trend of upper secondary education. The one that begins below the others refers to tertiary education

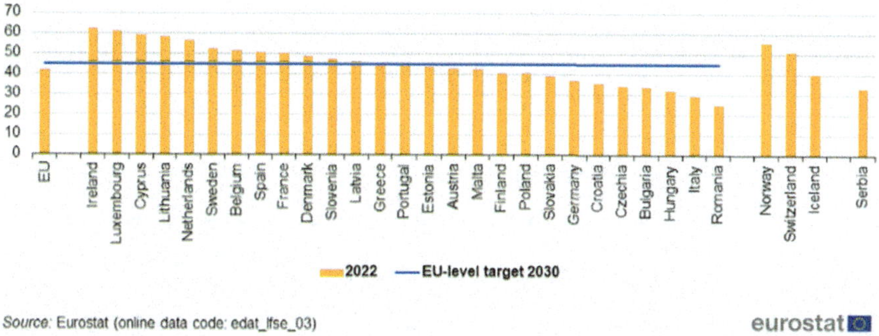

Fig. 5.14 Population aged 25–34 years with tertiary educational attainment (ISCED 5–8), 2022 (% of the population aged 25–34 years). *Source* Eurostat (2023)

education has progressed for people aged 25–54, with respect to older people. In fact, in the EU the average section of the population had an average of 37.7% of people having a high level of education, against an average of 22.7% for older people.

In 2022, among younger people (those aged 25–34), more females had the highest level of education, with more than the 45% EU target since 2019—now about 47–48%, against a little more than 36% for males in 2022 (see Fig. 5.15).

At the same time, also the upper secondary education level of the younger part of the population has increased in the last decade in all countries, with the exception of Austria, Germany and Sweden, as shown by Fig. 5.16 It is a little surprising that this level is highest in some of the states of the former Yugoslavia and Greece, in addition to Ireland and Lithuania.

All in all, we can say that, while the education level of the EU population has increased in the last years, there are countries that are still below the average level of 2012, not only in the South of Europe (Spain), but also elsewhere, where the average education level of the whole population is higher (Luxembourg, Denmark and Germany).

5.2.4 Finding a Job

Employment rates across OECD countries are higher for people having higher educational attainments. In fact, in 2021, on average, employment rates for tertiary-educated young adults (25–34 years-old) were 8 p.p. higher than those of people of the same age having attained upper secondary or post-secondary non-tertiary education and 26 p.p. higher than those of young adults having below upper secondary education. At the same time, higher educational attainment is negatively correlated with lower rates of unemployment and labour-market inactivity.

As partly seen before, tertiary attainment among 25–34 years-old has increased over the past two decades, with no decline in its labour-market value. In 2021,

Table 5.8 Population aged 20–24 years with at least an upper secondary education (ISCED 3–8) in the various EU and other countries, 2022 (% of the population aged 20–24 years). Eurostat (2023)

	25–54 years					55–74 years				
	Low	Medium	of whcih		High	Low	Medium	of which		High
	(ISCED 0–2)	(ISCED 3–4)	General orientation	Vocational orientation	(ISCED 5–8)	(ISCED 0–2)	(ISCED 3–4)	General orientation	Vocational orientation	(ISCED 5–8)
EU	18.1	44.2	10.2	34.1	37.7	31.6	45.7	8.7	37.0	22.7
Belgium	14.5	36.5	10.8	25.7	49.0	31.3	36.1	13.0	23.1	32.6
Bulgaria	15.8	52.6	24.7	27.9	31.6	20.5	57.8	24.1	33.7	21.7
Czechia	5.4	65.7	5.5	60.3	28.8	9.0	75.1	4.9	70.3	15.9
Denmark	16.5	37.4	7.9	29.5	46.1	25.9	44.3	4.4	39.8	29.8
Germany	17.0	49.0	5.4	43.6	34.1	17.1	55.2	2.2	53.0	27.7
Estonia	11.3	45.9	17.0	28.9	42.8	10.7	51.0	19.4	31.6	38.3
Ireland	8.8	33.3	16.9	14.8	57.9	31.0	35.9	20.7	13.4	33.1
Greece	14.5	47.9	27.1	20.9	37.6	41.4	34.4	25.2	9.2	24.3
Spain	32.1	23.1	12.8	10.3	44.8	53.7	19.8	12.6	7.2	26.5
France	13.7	39.8	9.3	30.6	46.5	30.9	43.5	10.7	32.8	25.6
Croatia	8.6	62.7	3.5	59.1	28.7	25.0	58.0	3.3	54.7	17.0
Italy	31.7	44.9	10.9	34.0	23.4	56.2	32.5	8.1	24.4	11.3
Cyprus	12.2	35.0	23.4	11.6	52.9	32.3	39.4	28.9	10.4	28.3
Latvia	9.4	47.8	20.9	27.0	42.8	6.5	64.6	22.0	42.6	28.9
Lithuania	6.5	42.0	16.5	255	51.5	3.5	64.3	22.4	41.9	32.2
Luxembourg	15.7	27.4	:	27.4	56.9	31.0	36.2	:	36.2	32.8
Hungary	12.5	56.1	13.0	43.1	31.4	17.4	61.3	9.6	51.7	21.3

(continued)

Table 5.8 (continued)

	25–54 years					55–74 years				
	Low	Medium	of whcih		High	Low	Medium	of which		High
	(ISCED 0–2)	(ISCED 3–4)	General orientation	Vocational orientation	(ISCED 5–8)	(ISCED 0–2)	(ISCED 3–4)	General orientation	Vocational orientation	(ISCED 5–8)
Malta	29.1	35.8	19.9	15.9	35.1	65.8	21.0	11.7	9.3	13.2
Netherlands	14.7	35.8	6.7	29.1	49.5	34.7	36.4	8.6	27.8	28.8
Austria	12.5	48.4	6.2	42.1	39.1	21.5	54.5	4.7	49.9	23.9
Poland	5.7	55.5	10.0	45.5	38.8	13.2	70.4	7.0	63.4	16.4
Portugal	31.4	32.4	20.1	12.3	36.2	68.6	14.9	12.4	2.5	16.5
Romania	17.5	60.4	8.1	52.3	22.1	294	62.0	6.2	55.8	8.7
Slovenia	6.7	48.7	5.2	43.5	44.5	18.6	56.7	3.2	53.5	248
Slovakia	5.6	62.0	41	57.9	32.3	11.7	70.8	4.4	66.4	17.6
Finland	9.8	45.9	8.7	37.2	444	21.4	44.5	3.9	40.6	34.1
Sweden	10.6	36.7	14.1	22.7	52.7	18.8	47.1	17.3	29.8	34.1
Iceland	17.8	36.6	16.1	18.7	45.6	23.3	46.2	11.7	33.5	30.5
Norway	16.1	32.4	9.4	23.0	51.5	20.5	46.0	11.2	34.8	33.5
Switzerland	127	39.2	6.9	32.3	48.1	18.0	49.8	6.8	43.0	32.2
Serbia	13.7	58.5	4.1	54.4	27.9	31.2	52.2	4.2	48.0	16.7

Source Eurostat (2023)
Note Data about general orientation is not available in Luxembourg due to very small sample size

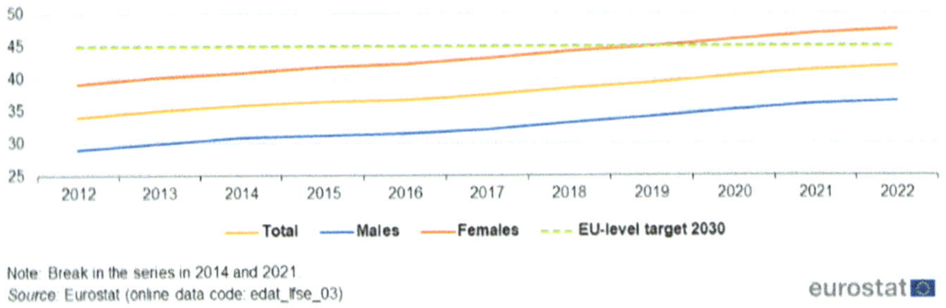

Fig. 5.15 Trend of population aged 25–34 with tertiary education attainment, by sex, %, EU, 2012–2022. *Source* Eurostat, Educational attainment statistics. *Note* The curve for females is always above the other continuous ones; that of males is always below them. The dotted line indicates the EU-level target for 2030

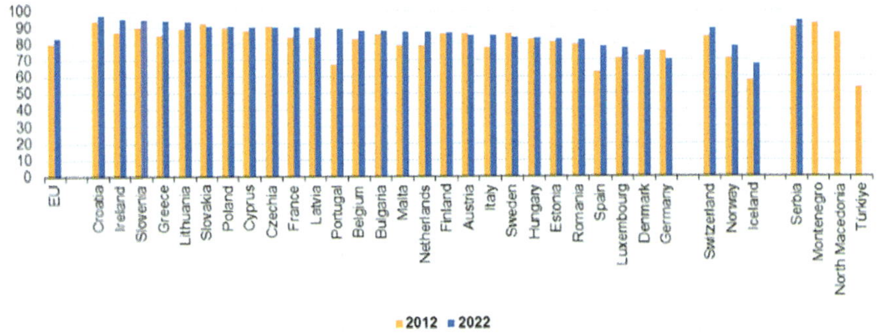

Fig. 5.16 EU and other countries' population aged 20–24 with at least an upper secondary education level (ISCED 3–8), 2012 and 2022 (% of the population aged 20–24). *Source* Eurostat, Educational attainment statistics

across OECD countries, the average unemployment rate of tertiary-educated younger adults was 2 p.p. below that of their peers with upper secondary or post-secondary non-tertiary attainment and 9 p.p. lower than for those with below upper secondary attainment, with almost no change from the year 2000 (OECD, 2022: 64).

Figure 5.17 below shows that in 2016 54% of young people in the EU did not work while studying for their highest completed qualification, the largest rate of non-working students being in Romania (91%) and the smallest in Finland (8%). In this year, the percentages were everywhere lower than in 2009, with the exception of Denmark, Austria, Ireland, Cyprus, Bulgaria and Greece.

In 2017, the European Commission, together with the European Parliament and the Council, introduced the European Pillar of Social Rights Action Plan. This aimed at promoting fair and efficient labour markets and welfare systems across Europe. In order to achieve this goal, the EU set three targets for 2030. One of these targets was attaining a 78% employment rate. In 2022, people aged 20–64 years employed in the EU were a total of 193,458,000, i.e., 74.6% of all people in the corresponding

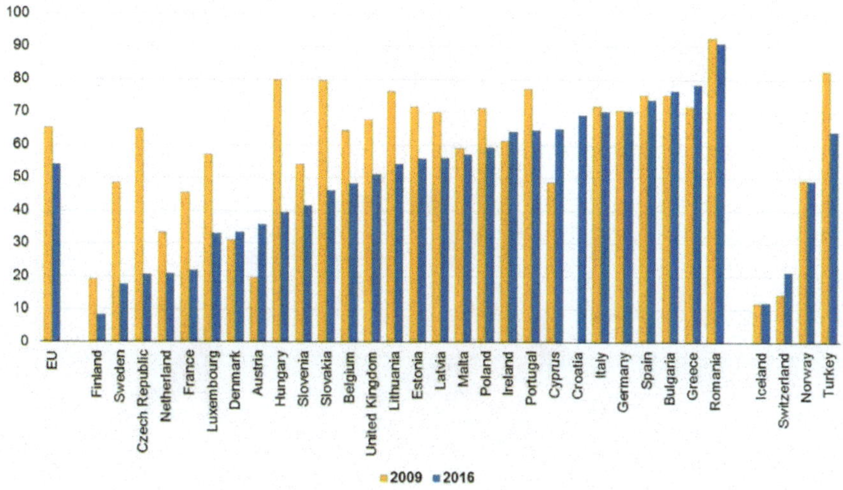

Fig. 5.17 Fraction of people aged 15–34 with some form of education who did not work while studying for their highest qualification, 2009 and 2016 (% of people aged 15–34 who had completed any formal education). *Source* Eurostat, Statistics explained, https://ec.europa.eu/eurostat/statistics-explained/index.php?title=Young_people_on_the_labour_market_-_statistics

age group. 11 EU countries reported attainment of this target already in 2022, while 11 countries were in the range of 70–78%, with the remaining 5 countries having employment rates below 70% (among them: Croatia, 69.7%; Spain, 69.5%; Romania, 68.5%; Greece, 66.3%; Italy, 64.8%). On the opposite side, 5 EU countries had an employment rate higher than 81%: the Netherlands (82.9%), Sweden (82.2%), Estonia (81.9%), Czechia (81.3%), Malta (81.1%) (see Fig. 5.18).

As can be seen from Fig. 5.19, in 2021 the employment percentage among people aged 25 to 64 having an education level below upper secondary schools was below the OECD average (58%) in many European countries (Belgium, where it was as low as 45%, Poland, Slovenia, Italy, Ireland, France, Greece, Finland and Austria, with 54%).

In 2020, the rate of employment of people aged 55–64 was rather lower than that of those aged 20–54 (59.6% against 76.2%). This depended on their educational attainment level, as the higher their education level, the higher their employment rate, which reached 75.6% and 85.7% for people of the two fractions of age having a high level of educational attainment (see Fig. 5.20).

The unemployment rates too are highly sensitive to the education level, which means that they are higher for lower education levels. We can see this by comparing the values indicated by the two following figures. In fact, Fig. 5.21 says that the average unemployment level of OECD people having below upper secondary education level was a little higher than 10%, whereas that of people with tertiary level education was only 4%. Figure 5.22 shows that unemployment of people having

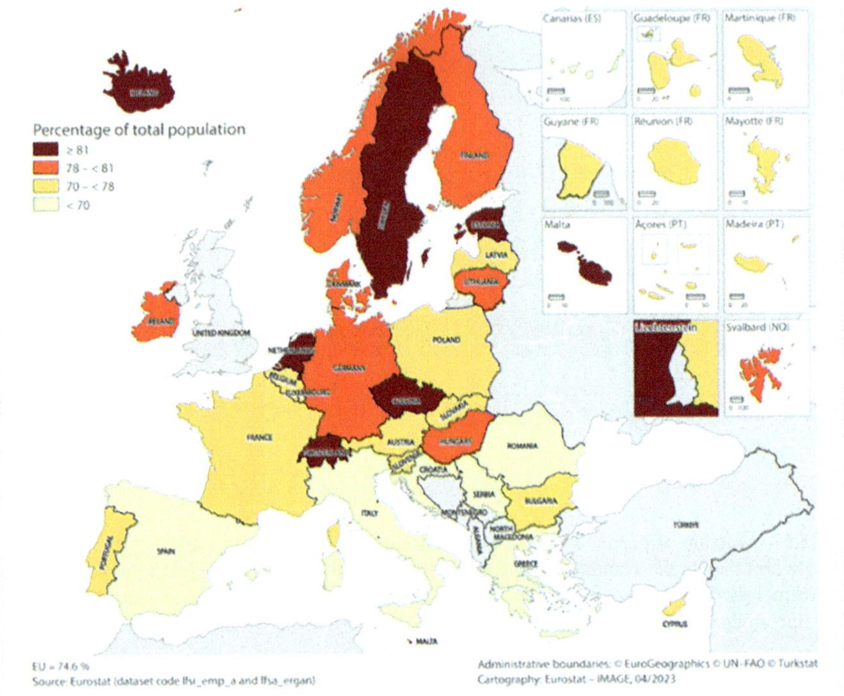

Fig. 5.18 Employment rate of different European countries, compared with the EU 2030 target, 2022. *Source* Eurostat, Employment—annual statistics

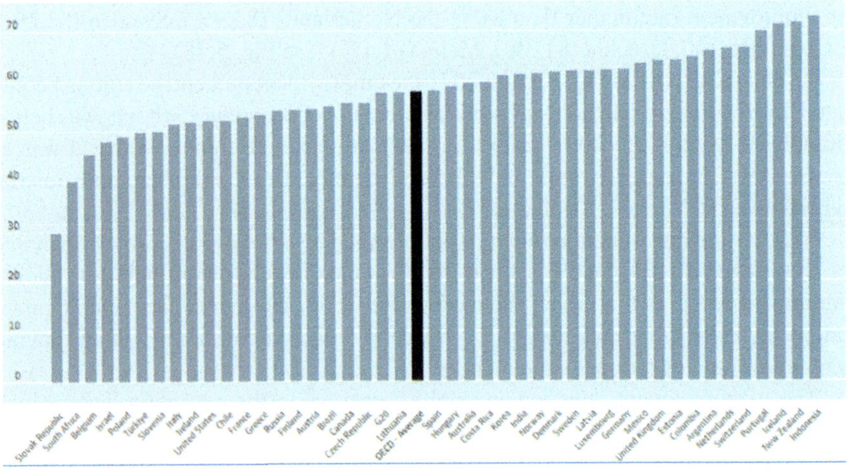

Fig. 5.19 Employment by education level below upper secondary schools, % of 25–64 years-old, 2021 or latest available. *Source* OECD (2022)

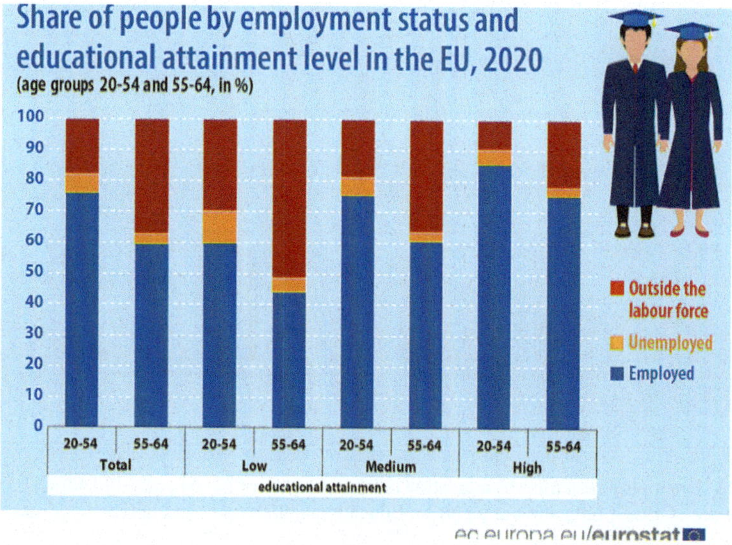

Fig. 5.20 Share of people by employment status and educational level, EU, 2020 *Source* Eurostat. *Note* The histograms indicate, from below: employed, underemployed, outside the labour force

attained a tertiary level education is much lower than that of people with a lower level.

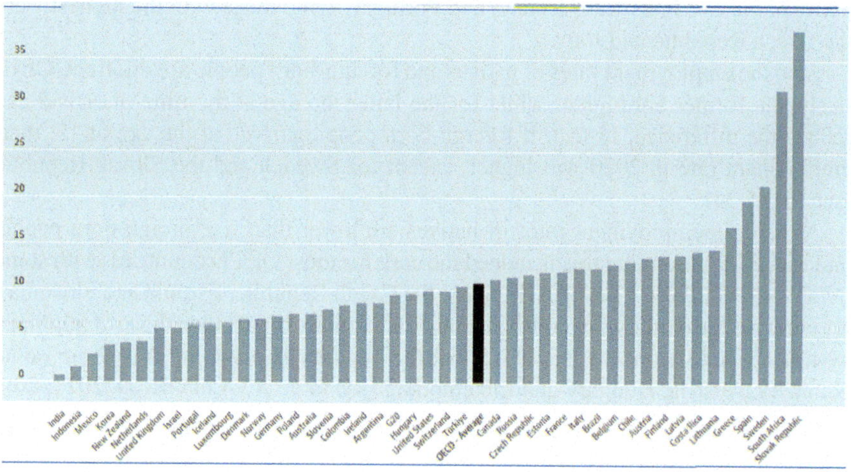

Fig. 5.21 Unemployment rates by education level, below upper secondary, % of 25–64 years-old, OECD countries, 2021. *Source* OECD, Education at a glance: Educational attainment and labour-force status

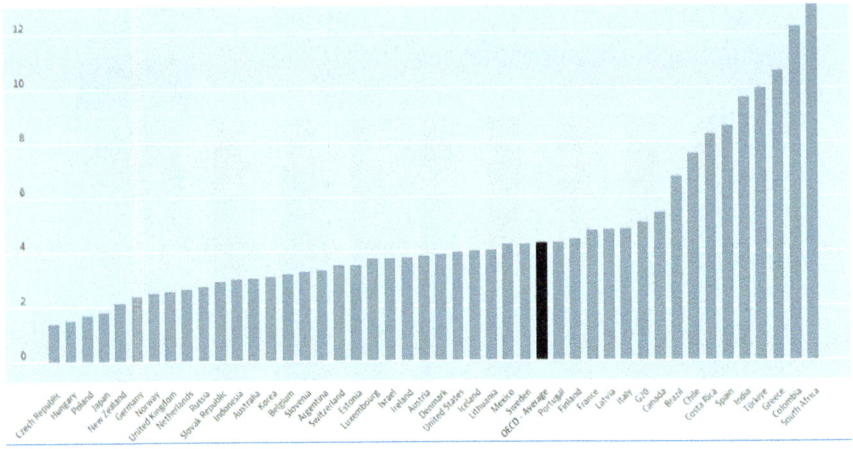

Fig. 5.22 Unemployment rates by education level, tertiary, % of 25–64 years-old, OECD countries, 2021. *Source* OECD, Education at a glance: Educational attainment and labour-force status

Obviously, also average employment rates are different for people with different education. In fact, they are 61% for people with below secondary education, in contrast to 75% for people having tertiary education (OECD, 2021).

Figure 5.23 opens another perspective, that of employment by gender, and shows, as expected, that employment rates in 2020 were higher for men than for women, with a difference of about 25 p.p. The European countries with the lowest difference between the two rates were Iceland and Portugal, while those with the highest were the Czech Republic and Italy.

Also the employment rates of natives and foreign-born people are different. Obviously, the former are higher, while for the latter the age at the time of arrival can reduce the difference. In fact, if foreign-born people arrived at the age of 15, their employment rate in 2020 was higher, except for Sweden and the Slovak Republic (see Fig. 5.24).

As said, unemployment rates of natives are lower than for foreign-born people and Fig. 5.25 shows that this is indeed the case for most OECD countries, with some notable exceptions (in Europe, Hungary, the Czech Republic, Estonia and Slovenia). By contrast, the employment rates of natives are usually higher than those of migrants, even if arrived before the age of 15, but the contrary happens in Europe for some countries, Austria, Hungary and Luxembourg (see table A3.4 in OECD, 2021: 80).

5.2.5 *Social Attainments and Education*

In this and the next sub-section, we want to draw some provisional conclusions from the analysis contained in the previous sub-sections and chapters.

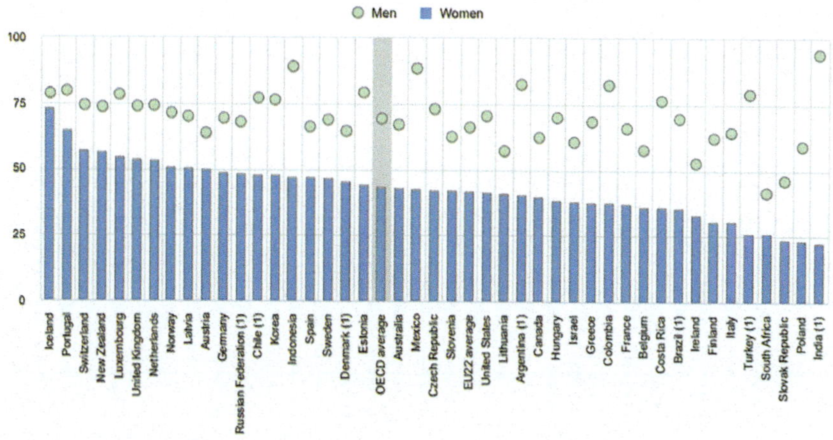

1. Year of reference differs from 2020. Refer to Education at a Glance Database for details.
Countries are ranked in descending order of the employment rate of 25-34 year-old women with below upper secondary attainment in 2020.
Source: OECD (2021), Education at a Glance Database, http://stats.oecd.org. See *Source* section for more information and Annex 3 for notes (https://www.oecd.org/education/education-at-a-glance/EAG2021_Annex3_ChapterA.pdf).

Fig. 5.23 Employment rates of 25–34 years-old with below upper secondary attainment, by gender (2020), %, OECD countries. *Source* OECD (2021)

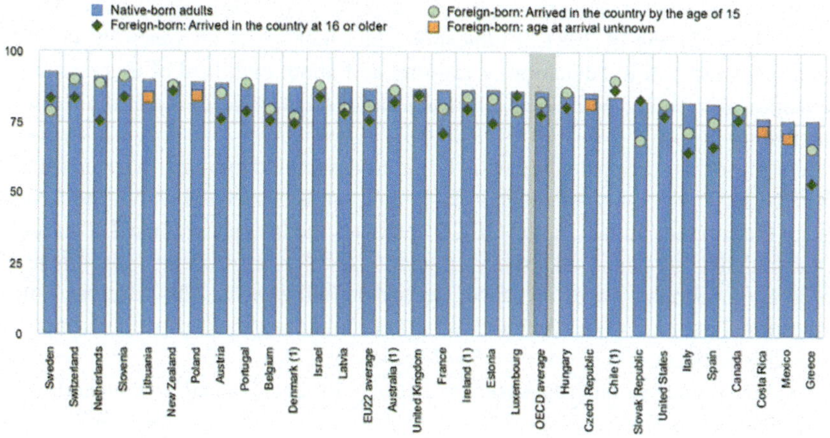

1. Year of reference differs from 2020. Refer to the source table for more details.
Countries are ranked in descending order of the employment rate of tertiary-educated native-born adults.
Source: OECD (2021), Table A3.4. See *Source* section for more information and Annex 3 for notes (https://www.oecd.org/education/education-at-a-glance/EAG2021_Annex3_ChapterA.pdf).

Fig. 5.24 Employment rates of native- and foreign-born 25–64 years-old with tertiary attainment, by age at arrival in the country (2020). *Source* OECD (2021)

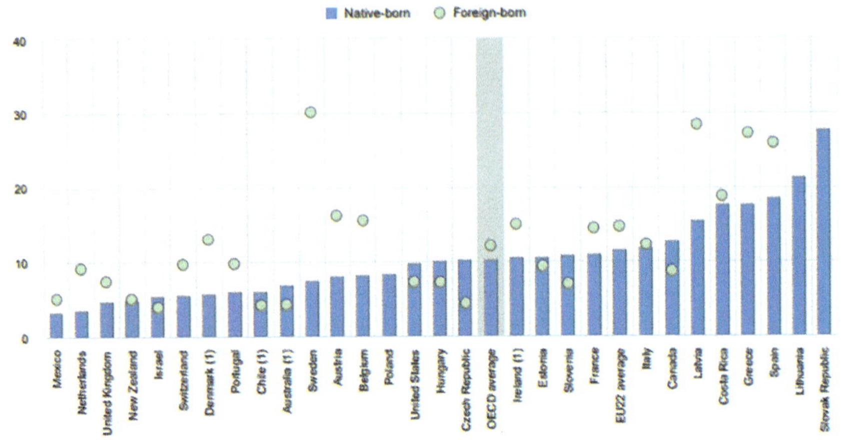

1. Year of reference differs from 2020. Refer to Education at a Glance Database for details.
Countries are ranked in ascending order of the unemployment rate of 25-64 year-olds native-born adults with below upper secondary attainment.
Source: OECD (2021), refer to Education at a Glance Database, http://stats.oecd.org. See Source section for more information and Annex 3 for notes (https://www.oecd.org/education/education-at-a-glance/EAG2021_Annex3_ChapterA.pdf).

Fig. 5.25 Unemployment rates of 25–64 years-old with below upper secondary attainment, by migration status (2020), %. *Source* OECD (2021)

At the beginning of 2021, a little more than half of 18–24 years-old in OECD countries were still in education or training and about one-third of them combined education or training with some form of employment, in a few cases linked to the curriculum.

We begin this section dealing with the habits associated to education. According to OECD (2022: 98), a higher educational attainment is associated with more frequent use of the Internet. In fact, in the OECD countries the Internet was used at least once a week in 2021 by nearly all tertiary-educated 25–54 year-old people, in comparison to 85% of people with below upper secondary attainment. The difference in the educational attainment between the two classes of education is especially important also with respect to those who are 55–74 years-old.

In some countries, such as Australia, Estonia, France, Iceland, Ireland and Latvia, only the students with tertiary-educated mothers expressed above the OECD average curiosity about other cultures, countries and economies. By contrast, for these students Italy was the only country with below-average interest of this kind. In this country there was almost no difference on this matter with respect to the mother's educational attainment. Similarly, students with tertiary-educated mothers in the OECD showed more positive attitudes towards immigrants and a stronger sense of global mindedness (OECD, 2022).

A low economic, social and cultural status of students' families requires an additional effort by the students. In fact, it has been ascertained that this status needs to be accounted when evaluating the effectiveness of education systems. Sometimes it is this low status—and not the nature of the system and of teaching per se—that leads to bad outcomes. As a consequence, proper public policies are needed to reduce disparities among students as to their socioeconomic background more than proper educational systems.

Investing public resources in education can reduce unemployment and favour future economic growth, by building the necessary skills. Public intervention can also offer incentives for hiring young people.

An important issue to deal with is that of the gender performance. According to van Hek et al. (2016) women have now reduced their subordination to men in education, possibly due to the larger labour market participation of the latter. There are indeed strong indications that now women outperform men in educational attainment in many countries, even if there are large differences among the various countries. However, UNESCO notes that, even if much progress has been made, large gender gaps still exist in many settings, most often at the expense of girls. An important aspect of female improved participation to education is that globally, against only 118.5 million girls, 125.5 million boys are out of school. However, women still account for almost two-thirds of all adults unable to read (UNESCO, Priority Gender Equality, https://www.unesco.org/en/gender-equality/education).

A second issue to discuss is the role of social classes for education. Education has been found as the strongest determinant of individuals' employment and chances of success, even more so in modern societies. Sullivan (2001) noted that children from middle class families gain educational credential, as they possess cultural capital, which is transmitted by their families. In addition to this, class differentials in material resources and educational aspirations account for other differences in educational attainment. According to Pensiero and Schoon (2019), education and its length have effects on the attainments of individuals, together with their income, social class and status. Notwithstanding their decline, educational inequalities are still large in recent cohorts, as origin still affects occupational outcomes, even for individuals having the same education (Triventi, 2018). Richardson et al. (2020) review the literature on the relationships between gender, ethnicity, and academic attainment in the United Kingdom higher education. In the last 15 years or so no much has changed as to inequalities in participation and attainment based on gender, social class, and ethnicity. Whitty (2001) noted the importance of the participation to education of the various classes, underlining the need of locating education policy within a broader social policy framework. Government policies had taken insufficient account of the conclusions of research on the impact of social class on educational success and failure. Social inclusion policies need to address middle-class self-exclusion from the mainstream provision of public schools together with the social exclusion of the working-class. Richardson et al. (2020) investigate social class inequalities in English school qualifications through pupils' outcomes for the GCSE. They find that there is a persistent social class separation, since pupils growing up in families of less advantaged social classes have less favourable school GCSE outcomes, with the consequence of limiting children's participation in more advanced education and restricting their options in the labour market. According to Hobbs (2016), there are large social class inequalities in educational achievement in the United Kingdom. In fact, more 'effective' schools are attended by children of a 'higher' social class.

Thus, social class differences have an impact on pupils' educational achievement. Caturianas et al. (2017) also find that education inequalities both affect and are affected by a number of aspects of the families of origin of students, such as their parental background.

Lagravinese et al. (2020) have considered the additional effort the students should do when they are coming from a disadvantaged economic, social and cultural status. In fact, some educational problems may not be due to the educational system per se, but to the economic, social and cultural gaps between students, determining a persistence of inequality of opportunities. Agić et al. (2016) find relations between certain characteristics of families—such as their completeness and the number of components, the level of education of parents, the choice of the school, availability of Internet access at home—and the average grade achieved by a student. With reference to the United States, Suárez-Orozco et al. (2015) analyse the educational situation of immigrants. These are among the least advantaged groups in the U.S. (and—usually—elsewhere) in terms of skills, education, and assets. Disparities existing before migration may change or not, according to the immigrants' interaction with a variety of post-migration situations (García Coll & Marks, 2012; Portes & Rumbaut, 2001; Suárez-Orozco et al., 2008).

Social and economic inequalities have important and long-lasting effects on children's cognitive and socio-emotional development as well as on educational outcomes (Grantham-McGregor et al., 2007; Shonkoff and Garner, 2012). Many inequalities combine, producing a negative impact on the ability of marginalised children to learn (Suárez-Orozco et al., 2015). Thus, the gap between advantaged and disadvantaged children widens over time (Shonkoff and Garner, 2012). Stopforth et al. (2020) find a positive correlation between the less favourable educational results of children and the less advantaged condition of their families.

According to Leathwood and Archer (2004), inequalities of class, race and gender persist in different places and conditions of education. Van Hek et al. (2016) analyse the role of gender in educational attainment for cohorts born between 1950 and 1982 across 33 countries. From the model they employ to test their hypotheses they derive that women outperform men in many countries, because of emancipatory contexts, even if in some countries they remain backward, due also to high levels of religiosity during their adolescence. Rupavath (2021) analyses various aspects of access to education for the tribal communities in India, underlining the wider livelihood options that a literate person can have, not confined to pursue occupations which largely entail manual labour. He underlines that governmental institutions need to be sensitised about the educational needs of tribal students.

The negative influence of a scarce level of education on the status of citizens has to be considered when advocating the meritocratic principle, according to which social inequalities, educational and status attainment should depend only on the level of achievement and ability of people rather than on factors such as their social origin.

A third issue refers to the effects of education. With respect to these, Vera-Toscano et al. (2017) note that education nurtures social outcomes and promotes participation to the social life, especially when education and training are undertaken in adult life. Schoon et al. (2010) find a direct association of childhood intelligence at the age of 11 with more liberal social attitudes at the age of 33, but this is partly mediated via educational qualifications and to some extent also via the occupational status. Thomson (2018) underlines the need to understand better the role of the socioeconomic background of students—if the function of education is not simply to reproduce inequalities—and how this effect is transmitted. Education affects several quality of life outcomes for individuals, such as: income, mortality risk, perceived health status, time spent with children in developmentally enriching activities (Edgerton et al., 2012), early parenthood, child nutrition, and child abuse (Alderman and Headey, 2017; Cullinan and Cawley, 2016; Vik et al., 2016).

5.2.6 The Role of Teachers

Teachers play many more roles than simply delivering lessons or handling a classroom. The best educators take up a multitude of roles, depending on the requirements of the students they teach. Teachers are superheroes of society and the roles listed above are just a few of the many roles they undertake.

Early learners and young students usually observe the behaviour and approach of their teachers to mirror their own actions. The advice given by teachers is taken seriously and very often the students turn to their teachers for guidance and advice. A senior teacher can also play the role of a mentor to junior teachers, imparting them practical wisdom related to their experience.

A number of suggestions have been offered on the different roles that can be performed by a teacher. As an example, Nola (2023) says that a teacher is at the same time: a controller of the activities of students and an assessor of these; a prompter, encouraging students to participate and providing advice and guidance about how students should proceed in an activity; a resource provider, a kind of walking resource center; a participant, taking part in some activities of students; a tutor, providing advice and guidance and helping students clarify their ideas and limit their tasks.

According to Steliou (2020), a teacher must be a mentor, an inspiring mediator and a resourceful and continuous learner, collaborating with fellow teachers working together with parents; a participant creating an environment that encourages learning; a collaborator of students; a participant conducive to a learning environment; a role model interpreting figures of wisdom and kindness; a great listener to what the students say; an eloquent speaker able to convey information accurately and efficiently; a planner/organiser; a creative thinker, encouraging students to be creative too; an assessor, able to understand the factors affecting the students' learning; an all-round educator that can enhance the learning experience of the student; a researcher and creator of new teaching materials to keep up with the latest research; a person able to deliver learners with one-to-one support; someone executing new teaching methods.

According to Wanders et al. (2021), the role of teachers has to be conjugated with that of parents and friends. The three groups of people can stimulate the civic participation of students, in order for them to be future active citizens. This participation is one of the primary goals of civic education and induces young adults in voting, partisanship, and other forms of political participation for the development of societal interest among other adolescents. Teachers can assess the extent to which—and how—differences among adolescents arising from their socioeconomic status can be compensated. Each of the three groups of people (parents, friends and teachers, but mostly teachers) can contribute to the development of societal interest among adolescents.

Eduk (2016) says that the roles of a teacher are: to deliver lessons effectively; to plan engaging and relevant lessons; to give regular feedback to learners; to support learners with their studies. His or her responsibilities are: keeping up to date with subject knowledge; identifying learners needs; planning and preparing all lessons and resources; assessing pupils' performance; adhering to legislation related to health and safety, equality and diversity and safeguarding; teaching without discriminating or stereotyping; maintaining the confidentiality of learner information; acting professionally and with integrity; avoiding to do something which is part of someone else's role; undertaking activities only as part of the teaching role.

Toropova et al. (2019) investigate the relations between the qualities of a teacher, the perceptions of instructional quality received by students, and the achievement of the students in mathematics. A significant positive relationship has emerged between coursework in mathematics and the levels achieved by students in this field, together with a similar positive relationship between student perceptions of instructional quality and students' achievement. The experience of teaching is positively related with student achievement for the first 19 years of teaching experience, but declines afterwards.

According to Bhattacharjee (2021), the characteristic attributes making the biggest difference between a good and an outstanding teacher are: being compassionate with students and passionate towards learning; having in-depth understanding capacity; being patient and a good follower of own words; being great communicator and great at putting efforts.

Croizet et al. (2019) document the role played by the educational system in the (re)production of social inequality by a process involving soft coercion. In particular, academic achievement is considered as revealing students' individual merit, whereas it only reaffirms the differences relating to family backgrounds playing in the classroom. They summarize the results of observations and interviews conducted in preschools as well as experiments carried out in primary and secondary education. This research unveils some of the dynamics through which education, as an institution, creates the conditions for the construction, reproduction, and legitimation of the stratification of society.

References

Agić, H., Osmanbegović, E., & Suljić, E. (2016). The impact of certain socio-demographic conditions on the students' success in learning. *Research in Pedagogy, 6*(2), 8–20.

Alderman, H., & Headey, D. D. (2017). How important is parental education for child nutrition? *World Development, 94*, 448–464.

Assmann, M.-L., & Broschinski, S. (2021). Mapping young NEETs across Europe: Exploring the institutional configurations promoting youth disengagement from education and employment. *Journal of Applied Youth Studies, 4*, 95–117.

Bhattacharjee, M. (2021). What is the role of a teacher in a student's life? *Edumilestones Blog.* https://www.edumilestones.com/blog/details/Role-of-a-Teacher-in-a-Students-Life

Brown, C., Douthwaite, A., Savvides, N., & Costas Batlle, I. (2022). Five mechanisms for tackling the risks to NEEThood: Introducing a pathway to change to guide educators' support strategies. *International Journal of Adolescence and Youth, 27*(1), 457–474.

Brunello, G., & De Paola, M. (2014). The costs of early school leaving in Europe. *IZA Journal of Labor Policy, 3*(22), 1–11.

Buitrago-Hernandez, P., Fuchs, A., & Cancho, C. (2019). *Exploring the diversity of young people not in employment, education or training (NEET): The gender profile of NEETs in Georgia and Armenia, developmental psychology.* The World Bank, April.

Caroleo, F. E., & Pastore, F. (2019). *The Italian low-growth conundrum: An assessment and some policy lessons.* CESifo Forum, ISSN 2190-717X, ifo Institut—Leibniz-Institut für Wirtschaftsforschungan der Universität München. München, 20(1), 33–39.

Caroleo, F. E., Rocca, A., Mazzocchi, P., & Quintano, C. (2020). Being NEET in Europe before and after the economic crisis: An analysis of the micro and macro determinants. *Social Indicators Research, 149*, 991–1024. https://doi.org/10.1007/s11205-020-02270-613

Caturianas, D., Užpelkienė, L., & Migliarini, V. (2017). *The role of education for social inequality in modern societies (with a special perspective on EU Member States).* NESET II Ad Hoc Question No. 5.

Coles, B., Hutton, S., Bradshaw, J., Craig, G., Godfrey, C., & Johnson, J. (2002). *Literature review of the costs of being "not in education, employment or training" at age 16–18.* Social Policy Research Unit, University of York, University of Hull, Queen's Printer.

Coles B., Godfrey, C., Keung, A., Parrot, S., Bradshaw, J. (2010). *Estimating the life-time cost of NEET: 16–18 year olds not in education, employment or training.* Research conducted for the Audit Commission by University of York, York.

Coll C. G., Marks A. K. (eds.) (2012). *The Immigrant Paradox in Children and adolescents: Is Becoming American a Developmental Risk?* American Psychological Association. https://doi.org/10.1037/13094-000

European Commission. (2022). *Education and training monitor 2022. Comparative report.* ec.europa.eu/education/monitor

Croizet, J.-C., Autin, F., Goudeau, S., Marot, M., & Millet, M. (2019). Education and social class: Highlighting how the educational system perpetuates social inequality. In: J. Jetten & K. Peters (Eds.), The social psychology of inequality. Springer International Publishing, pp.139–152. https://doi.org/10.1007/978-3-030-28856-3_9

De Luca, G., Mazzocchi, P., Quintano, C., & Rocca, A. (2020). Going behind the high rates of NEETs in Italy and Spain: The role of early school leavers. *Social Indicators Research, 151*, 345–363.

Edgerton, J. D., Roberts, L. W., & Peter, T. (2013). Disparities in academic achievement: Assessing the role of habitus and practice. Social Indicators Research, 114(2), 303–322. https://doi.org/10.1007/s11205-012-0147-0

Eduk. (2016). *The role, responsibilities and boundaries of a teacher.* Educating UK, 28th October. https://www.educatinguk.com/the-role-responsibilities-and-boundaries-of-a-teacher/

Eurofound. (2012). *NEETs—Young people not in employment, education or training: Characteristics, costs and policy responses in Europe.* Publications Office of the European Union.

European Commission, EACEA, & Eurydice. (2021). *The organisation of school time in Europe. Primary and general secondary education—2021/22.* Eurydice Facts and Figures. Publications Office of the European Union.

Eurostat. (2019). *Educational expenditure statistics*. Source: Eurostat. https://ec.europa.eu/eurostat/statistics-explained/index.php?title=Educational_expenditure_statistics

Eurostat. (2021). *Government expenditures on Education*. https://ec.europa.eu/eurostat/statistics-explained/SEPDF/cache/42735.pdf

Eurostat. (2023). *Educational attainment statistics*. https://ec.europa.eu/eurostat/statistics-explained/index.php?title=Educational_attainment_statistics

Goczek, Ł, Witkowska, E., & Witkowski, B. (2021). How does education quality affect economic growth? *Sustainability, 13*, 6437. https://doi.org/10.3390/su13116437

Grantham-McGregor, S., Cheung, Y. B., Cueto, S., Glewwe, P., Richter, L., & Strupp, B. (2007). Developmental potential in the first 5 years for children in developing countries. *The Lancet, 369*(9555), 60–70.

Helliwell, J. F., & Huang, H. (2011). *New measures of the costs of unemployment: Evidence from the subjective well-being of 2.3 million Americans*. NBER W. P. 16829, National Bureau for Economic Research.

Hobbs, G. (2016). Explaining social class inequalities in educational achievement in the UK: Quantifying the contribution of social class differences in school 'effectiveness.' *Oxford Review of Education, 42*(1), 16–35. https://doi.org/10.1080/03054985.2015.1128889

ILO. (2019). Young people not in employment, education or training. *Technical Brief*, N. 3. https://sustainabledevelopment.un.org/content/documents/26634NEET_Sida_brief.pdf

Lagravinese, R., Liberati, P., & Resce, G. (2020). The impact of economic, social and cultural conditions on educational attainments. *Journal of Policy Modeling, 42*(1), 112–132.

Leathwood, C., & Archer, L. (2004). Social class and educational inequalities: The local and the global. *Pedagogy, Culture and Society, 12*(1), 5–13. https://doi.org/10.1080/14681360400200186

Mattarelli, E., & De Santis, C. (2020). Educational system assessment: Italy and Finland. *Comparative Case Study*. https://doi.org/10.4995/HEAd20.2020.11076

Morrissey, T. W., Hutchison, L., & Winsler, A. (2014). *Family income, school attendance, and academic achievement in elementary school* (Vol. 50(3), pp. 741–753). https://doi.org/10.1596/31833

Mullis, I. V. S. (2019). *White paper on 50 years of NAEP use: Where NAEP has been and where it should go next*. American Institutes for Research.

Nola, A. (2023). *The seven roles of a teacher*. https://ng.opera.news/ng/en/education/f6cab3ae5dd78ef0cc874f6eae4cb6d7

OECD. (2006). *Education at a glance, OECD indicators 2006*. OECD Publishing. https://www.OECD-ilibrary.org/education/education-at-a-glance-2006_eag-2006-en

OECD. (2020c). *Education at a Glance 2020*. https://www.OECD-ilibrary.org/sites/5e4ecc25-en/index.html?itemId=/content/component/5e4ecc25-en

OECD. (2021). *Education at a Glance 2021: OECD indicators*. OECD Publishing. https://doi.org/10.1787/b35a14e5-en

OECD. (2022). *Education at a Glance 2022: OECD indicators*. OECD Publishing. https://doi.org/10.1787/3197152b-en

OECD. (2023). *Education spending (indicator)*. https://data.OECD.org/eduresource/education-spending.htm

Pensiero, N., & Schoon, I. (2019). Social inequalities in educational attainment: The changing impact of parents' social class, social status, education and family income, England 1986 and 2010. *Longitudinal and Life Course Studies, 10*(1), 87–108. https://doi.org/10.1332/175795919X15468755933380

Pileggi, A. (2018). L'INVALSI e la valutazione del sistema scolastico italiano. *Liberalismo Gobettiano*, 4 June.

Portes A., Rumbaut R. G. (2001). *Legacies: The Story of the Immigrant Second Generation*, University of California Press, Berkeley; Russell Sage Foundation, New York

Prince's Trust. (2010). *The cost of exclusion: Counting the cost of youth disadvantage in the UK*. Prince's Trust.

Quintano, C., Mazzocchi, P., & Rocca, A. (2018). The determinants of Italian NEETs and the effects of the economic crisis. *Genus, 74*(1), 1–24

Richardson, J. T. E., Mittelmeier, J., & Rienties, B. (2020). The role of gender, social class and ethnicity in participation and academic attainment in UK higher education: An update. *Oxford Review of Education, 46*(3), 346–362. https://doi.org/10.1080/03054985.2019.1702012

Rupavath, R. (2021). Poverty and education: Attainments and challenges for tribal communities. *Contemporary Voice of Dalit*, 1–18.

Schoon, I., Cheng, H., Gale, C. R., Batty, G. D., & Deary, I. J. (2010). Social status, cognitive ability, and educational attainment as predictors of liberal social attitudes and political trust. *Intelligence, 38*, 144–150.

Shonkoff, J. P., & Garner, A. S. (2012). The lifelong effects of early childhood adversity and toxic stress. *Pediatrics, 129*(1).

Steliou, A. (2020). 17 roles and responsibilities of a teacher. *Teacher Training*, March 30. https://cudoo.com/blog/14-roles-and-responsibilities-of-a-teacher/

Stopforth, S., Gayle, V., & Boeren, E. (2020). Parental social class and school GCSE outcomes: Two decades of evidence from UK household panel surveys. *Contemporary Social Science*. https://doi.org/10.1080/21582041.2020.1792967

Suárez-Orozco, C., Yoshikawa, H., & Tseng, V. (2015). *Intersecting inequalities: Research to reduce inequality for immigrant-origin children and youth.* William T. Grant Foundation Paper

Suárez-Orozco C., Suárez-Orozco M. M., Todorova I. (2008). *Learning in a New Land: Immigrant Students in American Society*, The Belknap Press of Harvard University Press, Cambridge, Massachusetts.

Sullivan, A. (2001). Cultural capital and educational attainment. *Sociology, 35*(4), 893–912.

Thomson, S. (2018). Achievement at school and socioeconomic background. An educational perspective. *Science of Learning, 3*(5), 1–2.

Toropova, A., Johansson, S., & Myrberg, E. (2019). The role of teacher characteristics for student achievement in mathematics and student perceptions of instructional quality. *Education Inquiry, 10*(4), 275–299.

Triventi, M. (2018). Social background, inequality of educational opportunities and occupational attainments. In *13rd Winter school on inequality and social welfare theory*, Alba di Canazei, 8–11 January 2018.

van Hek, M., Kraaykamp, G., & Wolbers, M. H. J. (2016). Comparing the gender gap in educational attainment: The impact of emancipatory contexts in 33 cohorts across 33 countries. *Educational Research and Evaluation, 22*(5–6), 260–282. https://doi.org/10.1080/13803611.2016.1256222

Vera-Toscano, E., Rodrigues, M., & Costa, P. (2017). Beyond educational attainment: The importance of skills and lifelong learning for social outcomes. Evidence for Europe from PIAAC. *European Journal of Education, 52*, 217–231.

Vik, F. N., Te Velde, S. J., Van Lippevelde, W., et al. (2016). Regular family breakfast was associated with children's overweight and parental education: Results from the ENERGY cross-sectional study. *Preventive Medicine, 91*, 197–203.

Wanders, F. H. K., Dijkstra, B., Maslowski, R., van der Veen, I., & Amnå, E. (2021). The role of teachers, parents, and friends in developing adolescents' societal interest. *Scandinavian Journal of Educational Research, 65*(5), 736–751.

Williams, E. (2023). What are SATs? And all your other questions answered everything you need to know about the SATs (UK). *Third Space*, June 6.

Chapter 6
An Ideal Education System

Introduction

This chapter tries to indicate the many features of an ideal education system, discussing their more or less acceptability. If the possible role of education in compensating for socio-economic disparities and the desirability of a European Education Area are uncontroversial, various can be the opinions on: the role of teachers and assessment methods, public or private provision, the optimality of various educational steps, the fields object of study and their more or less compulsory nature, the maximum number of students per class and homework schooling, the variety of assessment methods. The chapter closes with a discussion of the future of education.

6.1 The Role of Education in Compensating for Socio-economic Differences

We have seen—specifically in Sect. 5.2.5—that the socio-economic status of families has an influence on students' performance at school. Supportive environments and the quality and quantity of resources available are more frequently found in socio-economically advantaged schools, which amplifies rather than compensates for the socio-economic differences. This is due, first, to the usually higher literacy rate of these families, which can complement notions and methods taught at school. In addition, it can be explained by the possibility to choose private or selected public schools where students can be better taken care of. Avoiding discrimination deriving from the existence of private schools is not an easy task, as—on the one hand—freedom of education and the existence of private schools are usually guaranteed and—on the other—as a norm, teaching in public schools is imparted on equal standing to all students. Therefore, the issue that we must raise is how can compensating the differences that derive from socio-economic differences be done without recurring

N. Acocella, *Understanding Education in the European Union*,
SpringerBriefs in Education, https://doi.org/10.1007/978-3-031-62065-2_6

to a differential treatment of students that would add intolerable discriminations? In some countries, such as 'Estonia, Israel, Latvia, Macao (China), Montenegro, Norway and Tunisia, students in disadvantaged schools have access to learning environments and resources of comparable quality (or better, in some respects) as their peers in advantaged schools' (OECD, 2017: 2).

Nieuwelink et al. (2019) find that, particularly in secondary education, school exacerbates—instead of decreasing—social differences. About one-third of the variation in science performance across OECD countries is explained by the degree of equity in the allocation of educational resources across advantaged and disadvantaged schools, with more equitable systems performing better, on average.

OECD (2019: 106) rightly notes that it is necessary to avoid concentrating socio-economically disadvantaged students, in order not to negatively affect their education and ensure equity in education. Another reason to do that is to consider that a high concentration of disadvantaged students can lead to difficulties in attracting the most effective and experienced teachers, who are replaced by less-experienced teachers. On average across OECD countries, a higher number of teachers having at least a master's degree worked in advantaged schools; in addition, disadvantaged schools were more likely than advantaged schools to report that the 'school's capacity to provide instruction was hindered by a staff shortage teaching', or by 'a lack or inadequacy of educational material and physical infrastructure' (OECD, 2019: 106).

Education policies may partially compensate for this disadvantage, e.g., by providing more educational resources and staff to these schools. Another possibility is to encourage the best teachers to work in these schools, by offering them incentives. Finally, overtime schooling or home assistance could be offered for less-advantaged students.

According to UNESCO (https://learningportal.iiep.unesco.org/en/issue-briefs/improve-learning/socioeconomic-inequalities-and-learning) social and economic inequalities have impressive long-lasting effects on children's character and abilities, cognitive, socio-emotional development and educational outcomes (Grantham-McGregor et al., 2007; Shonkoff & Garner, 2012). In addition, 'multiple inequities combine, producing a negative impact on the ability of marginalised children to learn (Suárez-Orozco et al., 2015a, 2015b). Thus, the gap between advantaged and disadvantaged children widens over time (Shonkoff & Garner, 2012; Suárez-Orozco et al., 2015a, 2015b)' (UNESCO, https://learningportal.iiep.unesco.org/en/issue-briefs/improve-learning/socioeconomic-inequalities-and-learning). In contexts of high inequality, good quality and equitable education are key to the inclusive, peaceful, and sustained development of a society and a country. Ensuring the participation of marginalised groups and individuals in broader development processes helps reduce social inequalities. Inversely—unless attention is paid to equitable access, learning opportunities, and quality learning outcomes—'education can entrench existing inequalities or create new ones', according to UNESCO.

A problem similar to this arises for students who are resident in specific geographic contexts (e.g., the South of Italy, some regions of France such as the North of France and some parts of the Mediterranean coast or peripheries of towns). For instance, with reference to Italy, Pensiero et al. (2019) consider inequalities deriving not only

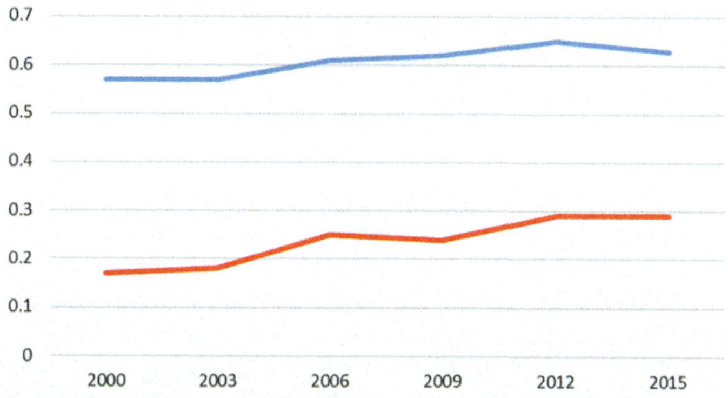

Fig. 6.1 Track attendance gap in upper secondary education by parents' ISCED (estimates obtained from PISA). *Source* Pensiero et al. (2019). *Note* The lower curve indicates lower ISCED; the upper curve indicates higher ISCED

from social origins, but also from the immigrant status and the area of residence and how they are intertwined. However, these authors find that the gap in the attendance rates of students at the age of 15 due to their parental education has narrowed from the beginning of the current century. In fact, in 2003 only less 18% of children from low-educated families (ISCED 1–2) attended the upper secondary track, whilst in 2015 they were 29% (see Fig. 6.1). Checchi and Peragine (2010) document the cumulative effects of lower social origins and the Southern origin, causing a brain drain from the South of Italy towards the Northern regions.

Similarly, the French population's standard of living is lower in the North of France and some parts of the Mediterranean coast, in the departments of Seine-Saint-Denis and Lot-et-Garonne. Between 2008 and 2011, the gap between the highest and lowest income levels widened in every department. Between 2012 and 2017, those gaps experienced a slight narrowing, under the general effect of a somewhat sharper rise in the lowest income levels. A strong reduction in inequality within rural and remote territories has also been documented in the following years (Kramarz et al., 2022). Due to fewer opportunities in their own region, students are often forced to move to Île-de-France (in the Paris area). This had 26.3% of all students enrolled in higher education in the whole country. More than 80% of students in Paris came from other regions in 2017 (Cathalat, 2020).

6.2 The Role of Teachers and Assessment Methods

Niemi and Jakku-Sihvonen (2011) say that high learning outcomes and high-quality teachers are elements of the education system in Finland. More specifically, the reason is that this country follows a policy of getting the right people to become teachers, developing them into effective instructors. Niemi and Jakku-Sihvonen clarify that in Finland all teachers are required to possess a master's degree. This qualifies teachers for teaching in primary and secondary schools. Instead, a kindergarten teacher is required to be a Bachelor in Education. Barber and Mourshed (2007) add that Finland's teachers are recruited from among the top 10% of each cohort. The corresponding value for Singapore and Hong Kong were 30%.

However, one has to recall that Finland scored 7th in reading and science and 16th in mathematics in the last PISA checks, whereas Singapore was 2nd in all fields and Hong-Kong scored 4th in mathematics and reading and 10th in science (see Table 5. 7 in Sect. 5.2.2). This is probably due to the fact that Finland, on one hand, and Singapore and Hong Kong have very different educational institutions.

Popova (2015) adds that teachers not only have a pedagogical education, but should also pass psychological tests on moral convictions and the attitude to children (teenagers) as well as to their job.

In addition to what we have already said, here we give a summary and a choice among the various suggestions that can be found in the literature—as well as our personal suggestions—about the role of teachers and of assessment methods.

In any case, teaching is a very delicate role requiring: specific training, updates and a close relation of teachers with students and their families.

Exchanges of experience abroad would be useful. The Erasmus program can play a role, not only for students, but also for teachers. In fact, the program 'Erasmus+' supports teaching assignments to schools abroad for professionals involved in pre-primary, primary and secondary school education.

Also the need arises of an adequate pay, in order to attract the best professionals to this role. In addition, the best teachers can be offered incentives in order to compensate them for operating in disadvantaged schools. This is one way to get rid of this disadvantage. In addition, more educational resources and staff could be provided for those schools.

Existing data about the performance of different class sizes, and student–teacher ratios in advantaged and disadvantaged schools offer contrasting indications, due to the fact that often there is a mix of lower class size and a higher proportion of disadvantaged students. Findings from PISA show, however, that, first, the observed small class size in disadvantaged schools does not fully compensate for the negative impact of the concentration of disadvantaged students within a school. In addition, in schools with a smaller number of students, teachers, on the one side, adapt their lessons to students' needs and knowledge—possibly changing the structure of the lesson when students find it difficult to follow; on the other side, in these schools teachers can provide individual help to student.

However, it must be considered that a higher number of teachers in schools with disadvantaged students can be ineffective, if the additional teachers are of a scarce quality, i.e., having neither experience—which makes them less confident in their ability to teach and to use a wide range of effective instruction approaches—nor a solid training. This has a negative effect on student learning (OECD, 2016, 2019). There have been increasing attempts to reduce attainment gaps due to socio-economic disadvantages by improving the quality of teaching, but the expected positive impact has been missing or in some cases even the contrary has happened (Riordan, 2022).

Assessment should be conceived as an integral part of education, as it clarifies to what extent the contents of teaching have met the goals of education, not only in learning the basics, but also in having acquired critical skills. Therefore, assessing a student's improvement can demonstrate his/her learning capacity.

Assessment and testing of teaching results have a strong impact on the lives and careers of both the students and their teachers. They have also an effect on the ways in which pupils are taught. Decisions taken within and by schools influence the prospects and opportunities of their pupils and of even greater importance are the results of national tests and examinations. If these results are used to pass judgments on teachers and schools, they also affect the ways in which pupils are taught. Because of this, it is essential that these results reflect and influence school learning in the best possible way. Finally, assessments can improve pupil achievement and summative assessments play an important role when seen with respect to national standards, by improving students' achievement (Wood-Wallace, 2016).

6.3 Public or Private Provision?

Before discussing the education policy in this field, we had better to recapitulate some facts. In OECD countries education is mainly publicly funded, but there is a substantial level of private funding at the pre-primary and tertiary levels. More specifically, at the tertiary level, less than 20% of all students are enrolled in private institution, which charge higher annual tuition fees than public institutions. By contrast, at least 80% of graduates in bachelor's, master's and master's long first degree programmes attend a public institution, but in most cases they have to pay a fee. In addition to funding public educational institutions, governments also finance private institutions, depending on the level of education. Private education is favoured by at least three factors: first, the often higher average class size of public schools (and, more generally, the teacher and educational material shortages of these schools) with respect to private schools; second, by the slightly lower ratios of students to teaching staff in private institutions than in public institutions at the lower secondary and upper secondary levels; third, by the more advantaged average socio-economic status of their families. Notwithstanding many shortcomings, students in public schools scored higher in reading than students in private schools, after accounting for students' and schools' socio-economic profile, on average across OECD countries (OECD, 2023).

As to the education policy, we have already indicated the current situation in most of the previous chapters and would like now to stress three things. The first one is that there have been various cases of privatisation of education, as shown by Table 6.1, not only in countries where the private supply of education has a long tradition (United States, Canada, Colombia and Brazil), but also in countries—such as the Scandinavian countries—where the public role has a long tradition.

The second notation is that existence of free and compulsory public schools is a question of democracy and equality. Philip et al. (2021) illustrate the cases of some developed and less-developed countries (Argentina, Egypt, Germany, Romania, and

Table 6.1 Paths towards privatisation in various countries. *Source* Verger et al. (2022)

Paths towards privatization		
Path	Privatization process	Paradigmatic cases
Reshaping the role of the state in education	Drastic advancement as part of a structural state reform adopted by neoliberal governments and consolidated by succeeding center-left administrations. Resulting in a quasi-market system where the state focuses on the regulation and monitoring	Chile. United Kingdom, New Zealand
Education privatization in social-democratic welfare states	Pro-market reforms, framed as part of a necessary modernization of the welfare state in a globalized context. Proactive leading or collaboration of social-democratic forces, usually less inclined to promote competition but to respond to middle-dass demand for diversification	Sweden, Denmark, Norway, Finland
Scaling-up privatization	Uneven but progressive alteration of the system through the authorization and encouragement of new forms of pnvate provision and management. School choice as guiding principle and with the key contribution of non-state actors forming advocacy coalitions	United States of America, Canada, Colombia, Brazil
De facto privatization in low-income countries	Expansion of a low-fee pnvate school sector, originally set by local *edupreneurs* responding to a growing education demand. Increasingly promoted by the international development community and supported by national governments as part of PPP schemes	Ghana, India, Kenya, Malawi, Nigeria, Pakistan, Peru, Dominican Republic, Jamaica

(continued)

Table 6.1 (continued)

Paths towards privatization

Path	Privatization process	Paradigmatic cases
Historical PPPs	Historical role and importance of private faith-based schools, incorporated In the state network during the expansion of basic in the twentieth century. Process mainly oriented to achieve a state-church compromise rather than on ideological grounds	Netherlands, Belgium, Spain, Argentina
Privatization by way of catastrophes	Advancement catalyzed by natural disasters or violent conflicts, which are framed by privatization advocates as an opportunity to reconstruct the system. Developed in contexts where the democratic deliberative process is undermined because the sense of urgency	New Orleans, El Salvador, Nicaragua Haiti, Guatemala, Honduras, Iraq

Source Verger et al. (2022)

Vietnam) and say that also in higher education the role of the state in giving financial support is central to provide access, equity, and funding for research. Other scholars have different opinions. In particular, according to these scholars, if private schools are opened, the state should not support any cost for them.

The third notation is that education certainly requires a high level of resources, but there is also the need of their good allocation in order to produce a highly skilled labour force and, more generally, to have a level of education high enough to enable citizens to take part to debates on social and political issues. Central government funding has the advantage of equity as it can ensure greater national uniformity of standards (Glover & Levačić, 2020).

Obviously, private education is expensive for students and therefore can be afforded only by rich people. In situations where public provision prevails or is exclusive, the co-presence of students belonging to higher-and lower- class families has the advantage of conveying some positive aspects for lower-status students. However, also upper-class students might derive an advantage from the presence of lower-class students. This would be a typical case of cross-fertilisation and a strong argument for making public provision of education prevailing or exclusive.

This statement opens the way for dealing with public–private partnership in education, which consists in 'any form of shared responsibility between government and private actors across the spheres of school finance, ownership, and management'. According to Baum (2018), a paper from which the previous citation is drawn, schools with this partnership outperform their public counterparts through sorting of more capable students. One can ask, however, whether this result depends also—or mainly—on the social composition of families sending their children to private schools (receiving public finance).

As to the applicability of the concept of public goods to the field of education from the basic to the higher level, doubts have been raised by Locatelli. As is well known, public goods are those goods that, independently from being produced by public authorities or not, are non-rival with respect to consumption of other goods and non-excludable, i.e., no one can be excluded from their consumption. Even if education is not a public good, the State should play the role of a regulator and provide and fund education, especially where this is not fully developed. Education is not a public, but certainly a common, good, for which greater public intervention and regulation are required, also in view of growing economic inequalities and of technological changes, in order to develop human beings and communities. From this point of view, one must consider that there is still no universal primary education (Locatelli, 2019).

Government provision of education is often insufficient. In some cases, low-cost private schools could achieve greater levels of quality provision than government schools, even if students' socio-economic background is accounted for, especially in some underdeveloped countries. After having reinforced the public provision of schooling—both in terms of access and quality—with the help of donors, governments should reduce remaining scarcity of school provision by promoting social dialogue and move towards acceptance also of low-cost private provisions (Pedró et al., 2015). Dinerstein et al. (2022) argue that new public schools increase public sector enrollment of students significantly, which leads to the closure of a large number of private schools, to lower prices asked by them and increased school quality by the surviving schools.

6.4 The Various Educational Steps

Apart from a few differences, many similarities exist in the various educational steps or levels of education. All European countries accept the levels of education established by UNESCO. These are the International Standard Classification of Education (ISCED) levels from 0 (referring to early childhood programmes) to 7 (referring to Master's or equivalent level programmes).

However, the ages to which these levels refer are different throughout the various European countries (see Fig. 6.2 as to the ISCED level 0).

The first difference refers to the role of kindergarten and compulsory education at that level. In Europe, compulsory education starts at various ages. Thus, in France and Hungary children are required to begin compulsory education at the age of 3. Attending at least one year of pre-primary education is compulsory in about half of the European education systems.

The second difference refers to models of primary and secondary education. In Sect. 4.1.1 we said that there are three main organisational models for all European

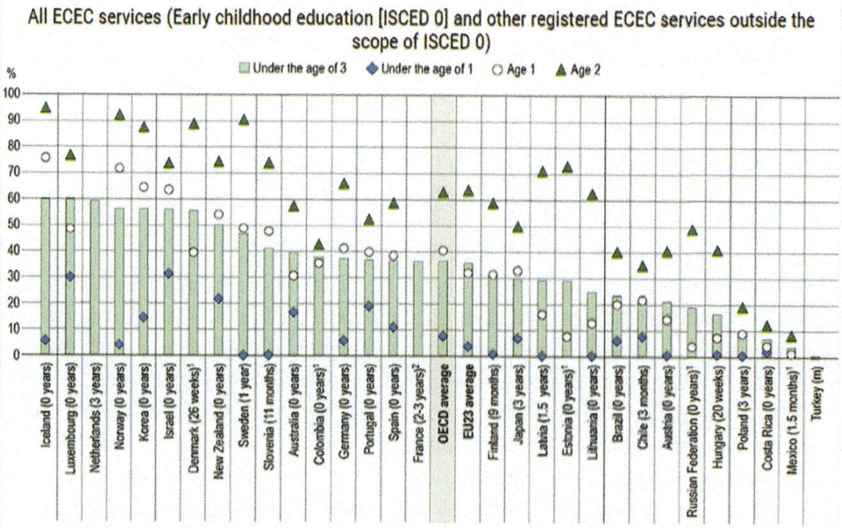

Note: Figures in parentheses refer to the age when ECEC systems start offering intentional education objectives.

1. Age 1 also includes children under the age of 1.

2. Data for other registered ECEC services come from the survey "*Modes de garde et d'accueil des jeunes enfants 2013*" conducted by the statistical division of the French Ministry for Solidarities and Health (DREES). Figures refer to the primary custody arrangements.

Countries are ranked in descending order of the enrolment rates in ISCED 0 of children under the age of 3.

Source: OECD (2019). Table B2.1. See *Source* section for more information and Annex 3 for notes

Fig. 6.2 Enrollment rates of children under the age of 3 in early childhood education and care, by age (2017). *Source* OECD (2019)

compulsory education systems: the Single structure education, the common core curriculum provision and differentiated lower secondary education. Apparently, no assessments about the preferability of one or the other model has been made until now, possibly due to the complexity of the issue when in presence of other differences between the countries.

Various grades are assigned in the different European countries to students to indicate their performance, especially in the upper secondary schools and at the university (see https://www.studyineurope.eu/grades).

6.5 The Fields Object of Study and Their Compulsory Nature

Knowledge taught and object of research is subdivided into disciplines. These are usually grouped into humanities (Arts, Languages, Philosophy, Anthropology, Law, History), scientific disciplines (Mathematics, Physics, Chemistry, Biology, Chemistry, Space Science, Rocket Science, Earth Science), and social sciences (Economics, Geography, Psychology, Sociology, Political Science).

Given this classification, there is certainly the need of strong compulsory basic knowledge in the various levels up to higher secondary schools. This basic knowledge certainly includes the native and some foreign language, mathematics and, possibly, history. The reason for teaching this group of disciplines is guaranteeing higher productivity in upper stages of education together with the possibility to easily move from one type of secondary school to another.

The native language and Mathematics are certainly indispensable basic knowledge, necessary for creative, logical and abstract reasoning, and essential to understand the concepts of all other disciplines.

Thus, the first goal at school should be to get children to love (or, at least, to know) words and numbers. Concerning the latter, the score—in the order—of the Netherlands, Ireland, Belgium, Denmark, Poland, Austria, the Czech Republic, Slovenia, Sweden, Germany, France, Spain, and Hungary—even if below that of Singapore, Japan, Korea and others—is above the OECD level in the PISA Programme for 2022. These countries are certainly well ahead other European countries, such as Greece, Hungary, Italy, and the Slovak Republic.

Similar, but with some differences, are the positions of the different European countries in PISA 2022 assessments for Reading and Science.

In some countries the basic disciplines taught at elementary and lower secondary schools include History, Geography and Science. However, one—as Redaelli (2020) has done—can object to this repetition. In particular, he says that 'reflective thinking only arises after the eleventh year, when adolescents are capable of making assumptions that are not necessarily related to concrete reality: they can uncouple their thinking from present times, ideate theories and elaborate thoughts which are not bound to their beliefs. Thus, there is likely no reason why disciplines such as History, Geography or Science should be studied before age eleven or twelve'. According to Redaelli, one could even question the existence of a system that comprises a primary and a middle school, due to the fact that the middle schools only last for three years, which compels teachers to tightly condense programs; in addition, at the end of the elementary courses students are forced to change their environment. This might require a general restructuring of the system, possibly unifying primary and lower secondary schools, as is the case in Germany or in the Netherlands.

The European Union has launched a program to face skills shortages—especially by women—in certain fields, in particular in the STEM fields—i.e. in Science, Technology, Engineering and Mathematics and Information and Communications Technology (ICT). These shortages are emerging due to the ongoing green and digital transitions of societies and the need to remain economically competitive at the global level. In addition, advanced 'transversal skills' are required that can be applied in particular in critical thinking and problem-solving and digital skills (European Commission, European Education Area. Quality education and training for all, https://education.ec.europa.eu/education-levels/higher-education/relevant-and-high-quality-higher-education).

6.6 The Maximum Number of Students per Class and Homework Schooling

The average number of students per class in OECD countries is more than 21 at the primary level, but in some countries such as Chile and China it is higher than 29, whereas it is around 15 in Luxembourg and the Russian Federation. In lower secondary education, the average number in OECD countries is about 23 students. The number is usually higher in public primary education schools, but the contrary happens in a number of countries at the secondary level.

From 2000 to 2010, the number of students per classroom has slightly decreased in most countries reporting these data, as can be seen in Fig. 6.3. But the size is still high, and this is due to the limited budget for education in many countries as well as to the difficulty of recruiting teachers, given their low salaries.

The class size is important, since usually a lower size facilitates teaching. However, in some cases there can be a good disciplinary climate and better teacher-student relations—to which the performance in reading in PISA 2009 is positively related—also in large classes (OECD, 2011). This may be the reason why a moderate negative correlation between the class size and achievement has been found in some countries (see, e.g., Koc & Çelik, 2015, for the case of Turkey).

A smaller number of students per class is not the only solution to improve learning. For example, another solution can be assigning support staff for crowded classes or

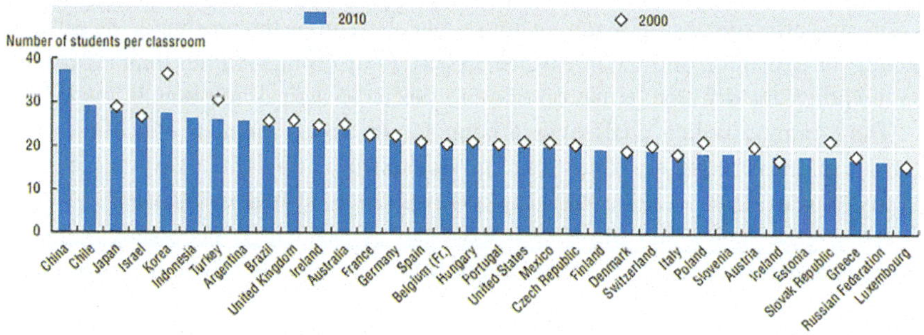

Source: OECD (2012), *Education at a Glance 2012*, Tables D2.1 and D2.5, available at: http://dx.doi.org/10.1787/888932667976 and http://dx.doi.org/10.1787/888932668071.

Fig. 6.3 Trends in average class size in primary education, 2000, 2010. *Source* OECD (2012)

improving teachers' training to handle big groups (García-Bulle, 2019). Adjunct teachers would be another solution. These should be recommended in particular when a limit to the number of students per class is feasible only with difficulties (e. g., due to indivisibilities).

Differences in disciplinary climate and teacher-student relations can help explain the differences in the overall average performance in reading between countries, such as in Albania and Japan, on the one side, which is better, and France, Greece and Spain, which is worse, on the other (OECD, 2020). This solution could be reached through a careful selection and update of teachers.

6.7 The Kinds of Assessments

Evaluation of teaching is not important in itself, but mainly in order to achieve improved student outcomes. Evaluation of students, teachers, schools and education systems can be done by various people. There can either be inspection and assessment within the schools or external services to evaluate teachers and/or schools.

Evaluation of schools consists in collecting, analysing, and interpreting information about any aspect of a program of education to judge schools' effectiveness, efficiency, and success (Msezane, 2020). External assessment is defined by UNESCO (https://www.ibe.unesco.org/en/glossary-curriculum-terminology/e/external-assessment) as 'a process and method of assessment developed and used by an examination body or agency other than the learner's school. This process commonly involves standardised testing, and often serves to grade candidates for further educational opportunities and/or for certification purposes'. Many countries test students at key points, such as is routinely done by teachers at the end of the year. In addition, there are national or international tests, such as INVALSI and FINEEC, in the first case, and PISA, in the second one (see sub-Sect. 5.2.2). The usefulness of the latter resides in the fact that they permit international comparisons and, possibly, improvements in the education system. One must, however, consider that the situation of education in the various countries is different not only because of the differential nature and effectiveness of teaching per se, but also for, on the one hand, the different models of education and the stress put on certain disciplines rather than others and, on the other, the different situation of countries as to their development level.

The common policy challenges arising for the assessment of education in the various countries refer to: educational governance and implementation; assessment of students; evaluation of teachers; school evaluation; system evaluation.

As to governance and implementation, the assessment is necessary mainly to ensure articulations within the assessment framework and to develop competencies for evaluation and for using feedback.

With reference to assessments of students, it is necessary to align them to educational standards and to balance these assessments with the assessments of teachers.

Evaluations of teachers are necessary for a number of reasons, mainly to rectify the assessment of students and to shape incentives for teachers.

External evaluation of the school is certainly needed and must take account of the different models of education and the stress put on certain fields rather than others. This evaluation is also necessary for comparing it with the internal one and possibly promoting some kind of incentive.[1]

Finally, in order to improve learning, system approaches that diagnose and address the causes at the root of low learning are needed. There are three systems of evaluation (Kaffenberger et al., 2022). For the first system, learning trajectories are required to evaluate the dynamics of children's performance. For the second system, a set of principles is needed for the evaluation and alignment of curricula and support of the teacher and the instruction systems. The third system of evaluation concerns a diagnostic framework and shows how the program takes a systems approach to improve learning.

6.8 The European Education Area

The Bologna Process is a series of ministerial agreements between European countries, initiated in 1999 after the signature by 29 European education ministers of the Bologna declaration aiming at the creation of a European Higher Education Area. The process is open to other countries in the European Cultural Convention of the Council of Europe (from Albania to the Vatican City).

In November 2022, the Progress Report prepared by the Commission highlighted the work done for the achievement of the European Education Area, and the remaining challenges, in order to fulfil the commitment to achieve it by 2025. The progress already made until now is testified by the following facts: almost 95% of children attend now early childhood education from the age of 4; more than 40% of young adults have acquired a higher education qualification; almost 90% of young people leave education with either an upper secondary diploma or enrolment in training; and, finally, 80% of recent Erasmus + graduates have gained employment in less than 3 months after graduating.

However, there is still further work to do in: abating the share of 15-years-old with low ability in reading, mathematics and science; increasing the participation of adults in learning; improving quality and equity in education and preparing teachers, trainers and school leaders; using digital education and promoting green education. In order to do that, EU member states should be supported in their reforms of education and training systems, enhancing synergies with other policy areas—such as research and innovation, social policy, employment and youth policies—as well as with EU funding instruments to better support national -reforms identifying targets and indicators to guide work and monitor progress (European Commission, no date).

[1] Internal evaluation includes the activities, inquiries and evaluations that schools use to assess the effectiveness of their program of self-review, planning and reporting processes.

Fig. 6.4 The six dimensions of the targets of the European education area. *Source* European Commission (2020)

The six dimensions of the targets of the European Education Area are: increasing quality and training in education, especially of young people; ensuring inclusion; guaranteeing gender equality; diffusing education and training policies and investments geared to the digital and green transitions by making well supported, highly competent and motivated teachers and trainers available; ensuring a closer and deeper cooperation between higher education institutions; guaranteeing a higher education cooperation as part of external policies (see more from Fig. 6.4).

In particular, demand for highly skilled people is increasing. It has been estimated that by 2025, high-level qualifications will be required by half of all jobs. To match this requirement higher education institutions are required to have a much deeper level of cooperation, fostering excellence, innovation and inclusion across Europe. A joint long-term strategy is required for having a systemic, structural and sustainable impact at all levels of their institutions. Students, staff and researchers will be mobile (in a physical, virtual or mixed form).

To monitor progress, the Commission proposed five EU-level targets, referring to:

- the share of low-achieving young people aged 15 in reading, mathematics and science, which should be less than 15%;
- the share of low-achieving eighth-graders in computer and information, to be less than 15%;

- children between 3 and the compulsory education age participating in early childhood education should be at least 98%, starting from a level of 90.3% in 2009;
- the share of people aged 20–24 years with at least an upper secondary qualification should be 90%, rising from the level of 2012;
- the share of people aged 30–34 with tertiary educational attainment should be at least 50%, rising from 32.6% in 2010 (European Commission, 2020).

The Commission also indicated six dimensions of the targets of the European Education Area, which are illustrated in the Fig. 6.4.

Table 6.2 shows the gap for both females and males between the current situation and the seven educational targets set for 2030 in the European Union. The situation is usually better for females, except for Vocational Education and Training (VET).

The distribution of underachievement in reading among the different European countries in 2015 and 2018 is indicated by Fig. 6.5. In 2018 this indicator was already lower than the target set for the whole European Economic Area in Estonia, Ireland, Finland and Poland. On the other hand, it was higher in 2018 than in 2015 in Sweden, Czech Republic, France, Lithuania, Hungary and Slovakia.

For mathematics and science, instead, the distribution of underachievement among the different European countries in 2018 is indicated by Figs. 6.6 and 6.7, respectively.

The underachievement gap between the low and high socio-economic status when combining reading, mathematics and science is usually lower than when looking at them separately.

In other terms, it is more likely for a 15-years-old to have a low score on one PISA scale than on all three PISA scales simultaneously. And, thus, 'across the EU, the shares of underachievement in reading (22.5%), mathematics (22.9%) and science (22.3%) are all higher than the share of underachievement in all three domains combined (13.0%). This illustrates how the new EU-level indicator on equity is built on a more severe definition of educational underachievement' (European Commission, 2022: 11).

Thus, 'the new EU-level indicator for equity in education is more favourable to some countries and less favourable to others. Some Member States compare more positively with the EU average, such as Czechia (in particular when compared in reading and maths separately), Slovenia (maths, reading) and Denmark (science). Other Member States compare more negatively with the EU average, such as Malta (in particular when compared to reading separately) and Cyprus (maths)' (European Commission, 2022: 11).

For this one can see Fig. 6.8.

Educational underperformance can be coupled with socio-economic status. In fact, 'students of low socioeconomic status are almost 6 times more likely to underachieve than students of high socioeconomic status' (European Union, 2022: 6). This direct relationship between students' performance in reading, mathematics and science and the socio-economic status of students' families is clear from Fig. 6.9. If we refer to Romania, in fact, 50% of students from families of a low socio-economic status

Table 6.2 Trends towards EU-level educational targets, 2022. *Source* European Commission (2022b)

Seven EU-level targets				
	EU-level target	Latest EU average		
		Total	Female	Male
1	By 2030, at (east 96% of children between 3 years old and the starting age for compulsory primary education should participate in early childhood education and care	93.0% [2020]	93.1 [2020]	93.0 [2020]
2	By 2030, the share of early leavers from education and training should be less than 9%	9.7% [2021]	7.9% [2021]	11.4% [2021]
3	By 2025, the share of recent VET graduates who benefit from exposure to work-based learning during their vocational education and training should be at (east 60%	60.7% [2021]	60.3% [2021]	61.0% [2021]
4	By 2030, the share of 25–34 year-olds with tertiary educational attainment should be at east 45%	41 2% [2021]	46.8% [2021]	35.7% [2021]
5	By 2025, at least 47% of adults aged 25–64 should have participated in learning over the previous 12 months	[*]	[*]	[']

(continued)

Table 6.2 (continued)

Seven EU-level targets				
	EU-level target	Latest EU average		
		Total	Female	Male
6	By 2030, the share of low-achieving 15-year-olds in reading, mathematics and science should be less than 15%	Reading: 22.5% [2018] Maths: 22.9% [2018] Science: 22.3% [2018]	Reading: 17.5% [2018] Maths: 22.9% [2018] Science: 21.2% [2018]	Reading: 27.4% [2018] Maths: 22.8% [2018] Science: 23.2% [2018]
7	By 2030, the share of low-achieving eight-graders in computer and information literacy should be less than 15%	[**]	[*]	[']

Source European Commission (2022b)
Source For target 1, Eurostat (UNESCO OECD Eurostat [UOE] joint data collection, for targets 2, 3, 4, 5, EU Labour Force Survey [LF5]); for target 6, OECD (Programme for the International Assessment of Students [PISA]); and for target 7, the International Association for the Evaluation cf Educational Achievement [IEA] (International Computer and Information Literacy Study [ICILS]).
Note [*] The underlying LFS data will be available biennially as of 2023[26]; ICILS 2023 data will be available in 2024[27]

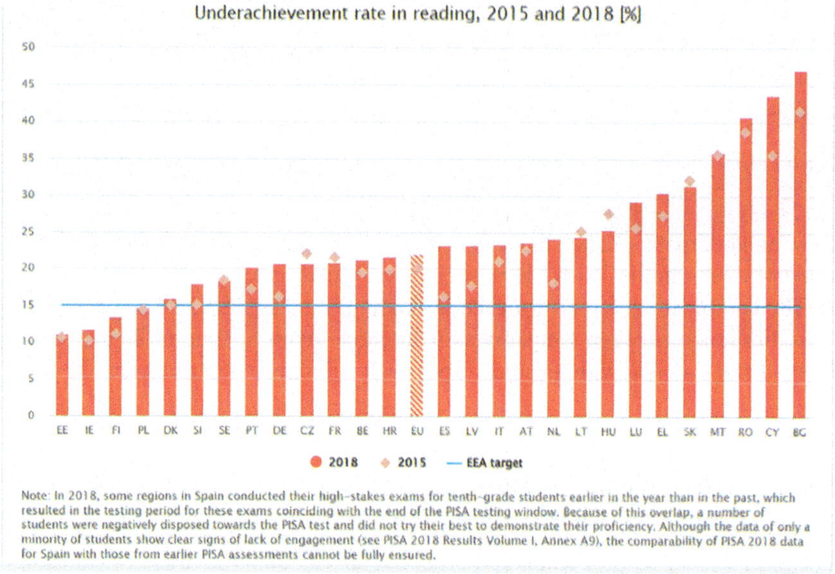

Fig. 6.5 Underachievement in reading in the different European countries, 2015 and 2018, %. *Source* European Commission (2019)

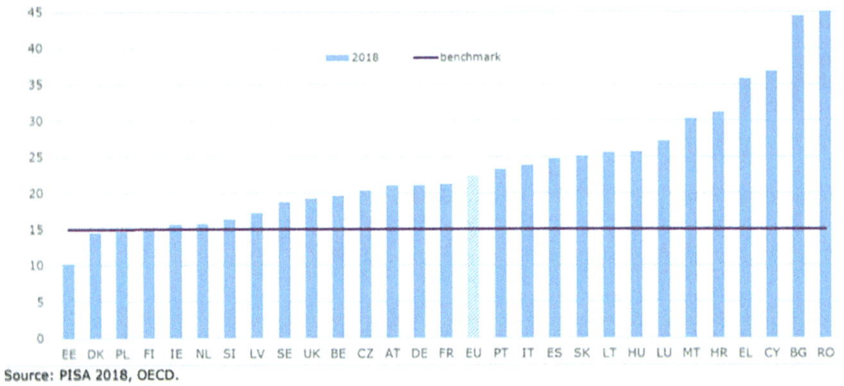

Fig. 6.6 Underachievement rate in mathematics in 2018 (%). *Source* European Commission (2019)

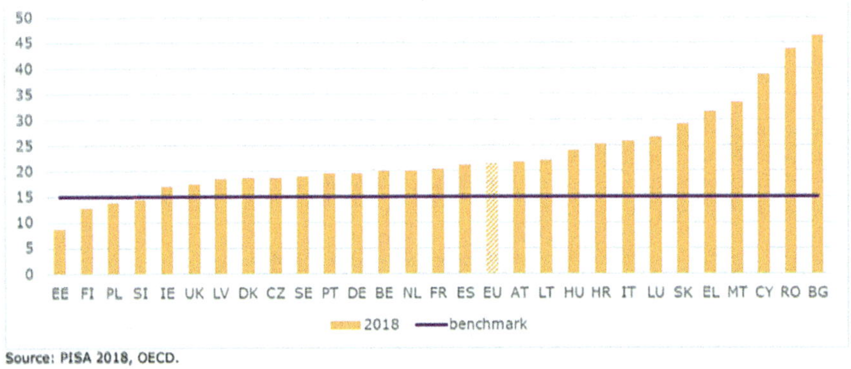

Fig. 6.7 Underachievement rate in science in 2018, %, European Commission (2019)

show an underperformance in these fields, as compared to a little more than 10% of students from families with a high socio-economic status.

The combined underperformance in reading, mathematics and science shows a high difference between Romania and Bulgaria—with a value of almost 40—and Finland and Estonia, with a value, respectively, of almost 10 and around 5.

6.9 The Future of Education

The 2021 Davos World Economic Forum has asked about the future of education in 20 years' time. An invitation has come from it to use imagination and rigour when thinking about the future, and to guard against the temptation to choose a favourite future, preparing for it alone. This is especially necessary because the world can change under shocks—like the pandemics—having different impacts on the various

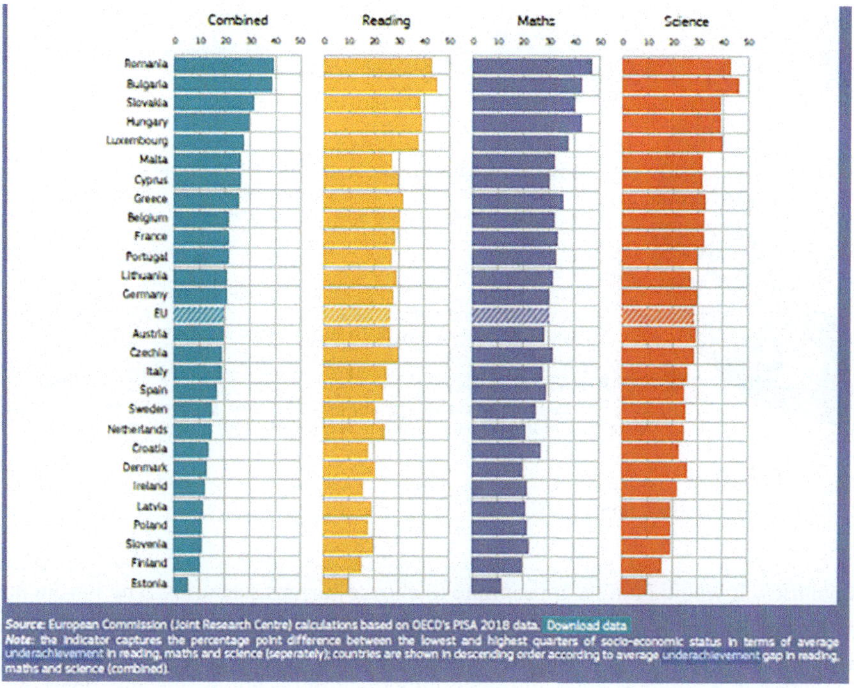

Fig. 6.8 Education performance in reading, mathematics, science and combined. *Source* European Commission and OECD (2019)

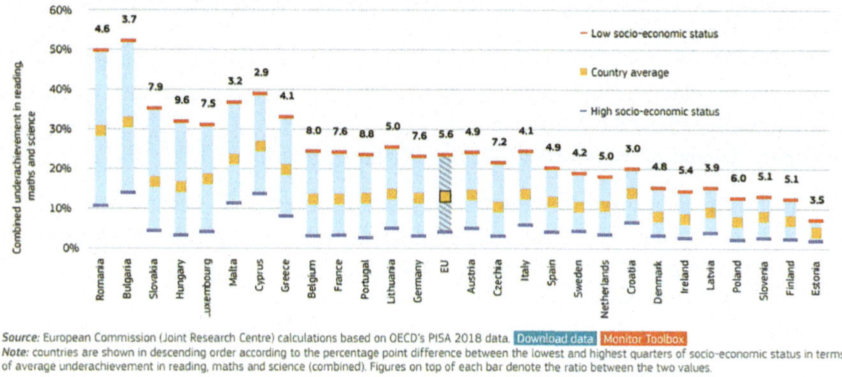

Fig. 6.9 Educational underperformance is coupled with socio-economic status. *Source* European Commission (2022)

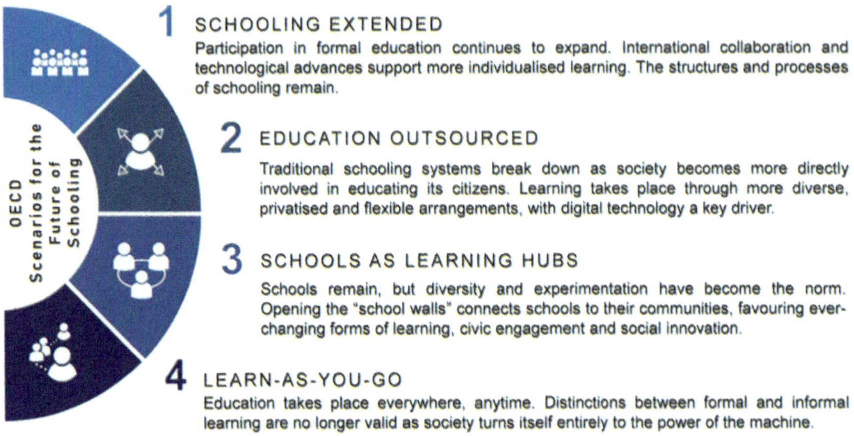

Fig. 6.10 The four possible future scenarios indicated at the 2021 Davos World Economic Forum. *Source* World Economic Forum (2021)

countries. It would then be appropriate to choose various scenarios. The Forum has indicated the four following scenarios (see Fig. 6.10), which could combine or take place each in a different place around the world.

Different scenarios have different important implications for the goals and governance of education, as well as the teaching workforce.

We can construct an endless range of such scenarios. The future could be any combination of them and is likely to look very different in different places around the world. Despite this, thinking about the future gives us the tools to explore the consequences for the goals and functions of education, for the organisations and structures, for the education workforce and for public policies. Ultimately, it makes us think harder about the future we want for education.

Sengeh D. and R. Winthrop (2022) make three remarks:

1. The transformation of the education system is urgent now, because the world is 'at an inflection point', due to the climate change, the changing nature of work, the increasing conflicts and authoritarianism.
2. The transformation of the education system must entail a fresh review of the goals of each system and their suitability for our times in so far as they can tackle inequalities and build resilience for a changing world.
3. Finally, each one of us (and our families, students, communities, etc.) should think of how the transformation of the education system can advance in his/her country or jurisdiction, considering: the need of developing a broadly shared vision and purpose; how to redesign the pedagogical core; how to position and align all components of the system in support of the pedagogical core and purpose.

According to eLearning Inside (2021), technological advancement in the twenty-first century has completely changed the way school works and has to prepare for the careers of future generations. The current approach to schooling is no longer a guarantee for a successful career. This is why we must ask for how to reform education to develop 21st-century skills for today's students. Students must be equipped with enhanced learning methods through a variety of approaches, such as: working in small groups; creating a modern learning environment, by encouraging project-based learning, visits to successful companies, learning about different cultures, etc.; developing talent critical thinking, creativity, empathy, and leadership skills; using digital technology skills.

6.10 Conclusions

In these conclusions we will extract the main aspects of the content of previous chapters, which allow us to deal with a number of issues, with specific reference to the European Union.

1. These are:
   - Why is education important?
   - The variety of European education systems.
   - Comparisons of the various education systems from different points of view.
   - Methods and models of teaching and attendance.
   - Is there a superior education model?
   - The need for accommodating different aspects of education.
2. As to the reasons why education is important, we can say that these are various:
   - Education is acquisition of pure knowledge, also when it offers scarce or no possibility of practical implementation of knowledge. This is in particular the case of issues related to astronomy, history, philosophy.
   - Apart from acquisition of pure knowledge, schools allow an exchange of experiences or the possibility to find applications of some of the notions imparted for reading, writing, geography, engineering, natural and applied sciences.
   - Some notions are purely instrumental, for the following three reasons: obtaining higher productivity, searching for a job and social upgrading.
   - The improvement in the productivity of the economic system derives from applications of notions, such as, more specifically, in the case of engineering, physics, chemistry, natural and applied sciences, economics, law.
   - People having higher education can more easily find a job and have a higher remuneration. In fact, in 2021 employment rates for tertiary-educated young adults were 8 p. p. higher than those having attained upper secondary, but not tertiary, education and 26 p.p. higher than young adults having only below upper education. At the same time, wages paid to workers having a higher level of education were also higher.
   - Education offers social upgrading, since literate people are usually more favourably considered by other people.

- Various devices are used in some countries to organise schools in a way to compensate for the socio-economic differences of students and, in any case, education per se tends to do so.
- We must say that all these positive effects of education, beyond being important per se, are instrumental for achieving equity, tolerance and democracy.

3. We found that a variety of European education systems exist, which are different from the points of view of:

   - Their organisation as to: division between the different grades and modality of passing from one to another grade; salaries of teachers.
   - Methods and models of teaching. The teaching methods are numerous: lecturing; demonstrating and debriefing; direct instructions and lecturing; online and multimedia learning; cooperative learning; inquiry-based learning; seminar and project-based learning. These methods are inspired by different theories and disciplines. As to models of teaching, which establish the modalities of transition between the different levels of education, there is also a variety, ranging from the Single structure to the Common core curriculum and Differentiated education.
   - The assessment of the learning capacities of pupils is an integral part of education, which has a strong impact on the lives and careers of both the students and their teachers, having also an effect on the ways in which pupils are taught.
   - European systems also differ as to the compulsory or not nature of education, from the point of view of both the age at which it starts and the compulsory nature of the fields object of study.
   - The various systems also differ as to the length of teaching, in terms of both the number of days of education and the hours of compulsory education per day.
   - As to the public or private nature of education provision, we must recall that there have been various cases of privatisation of school in Europe, but the various national education systems are mainly of a public nature. This is relevant as the existence of free and compulsory public schools is a question of democracy and equality.
   - The costs of private and public educational institutions are similar for the EU countries, but the costs of private institutions are usually (but not in Sweden) charged to the families of the students, whereas those of public schools are charged to various levels of governments. This has the effect of discouraging access of lower-medium income families to private education. The existence of free and compulsory public schools is a question of democracy and equality.
   - As to mobility, the Erasmus program provides funds for the mobility of university students and now is also available to secondary school students and teachers.

4. In discussing the existence of NEETs and their different size in the various European countries, we have concluded that this status derives from disaffection with school—certainly to be related to methods and models of teaching—in

addition to the difficulty of young people to find a job, which is different in the different European countries.

5. The final issue we have to indicate is whether there is a superior education model. This is a very difficult question to answer as the need arises for searching for a model that can accommodate different aspects of education as well as the different needs of the various countries.
6. As a (meagre) conclusion, we should say that, unfortunately, the future of education in the European Union is rather uncertain. There are differences as to: providers; models and methods of teaching and assessment; fields that are object of studies; role of teachers; proficiency; attendance and completion of studies; role of educations for compensating socio-economic differences. And these differences tend to persist. Therefore, there is still much work to do to implement the six dimensions of the European Education Area devised by the Bologna Process.

References

Barber, M., & Mourshed, M. (2007). *How the world's best performing school systems come out on top*. Mckinsey & Company.

Baum, D. R. (2018). The effectiveness and equity of public-private partnerships in education: A quasi-experimental evaluation of 17 countries. *Education Policy Analysis Archives, 26*(105). https://doi.org/10.14507/epaa.26.3436

Cathalat, A. C. (2020). *Higher education in France: Why territorial inequalities matter*, 26 May. https://www.right-to-education.org/blog/higher-education-france-why-territorial-inequalities-matter

Checchi, D., & Peragine, V. (2010). Inequality of opportunity in Italy. *Journal of Economic Inequalities, 8*, 429–450. https://doi.org/10.1007/s10888-009-9118-3

Dinerstein, M., Morales, D., Neilson, C., & Otero, S. (2022). *The equilibrium effects of public provision in education markets: Evidence from a public school expansion policy*, November 15. https://sebotero.github.io/papers/DNO_draft.pdf

European Commission. (2019). *PISA 2018 and the EU. Striving for social fairness through education.* https://education.ec.europa.eu/sites/default/files/document-library-docs/pisa-2018-eu_1.pdf

European Commission. (2020). *Achieving the EUROPEAN education area by 2025*. https://education.ec.europa.eu/sites/default/files/document-library-docs/eea-factsheet-sept2020_en.pdf

European Commission. (2022b). *Progress towards the achievement of the European education area.* Communication. file:///C:/Users/User/Downloads/progress%20towards%20the%20achievement%20of%20the%20european%20education-NC0422061ENN%20(1).pdf

European Commission (2022b). *Education and training monitor 2022*. Publications Office of the European Union.

European Union. (2022). *Education and training monitor 2022. Comparative report*. https://op.europa.eu/webpub/eac/education-and-training-monitor-2022/downloads/comparative-report/Education-and-Training-Monitor-Comparative-Report.pdf

eLearning Inside. (2021). *Modern education: How to prepare students for the jobs of the future*, April 16. https://news.elearninginside.com/modern-education-how-to-prepare-students-for-the-jobs-of-the-future/

García-Bulle, S. (2019). *Population and effectiveness in the classroom, how many students are too many? Observatory of the Institute for the Future of Education*, June 25.

Glover, D., & Levačić, R. (2020). *Funding education—Public and private systems, chapter 2 of Funding education*. UCL Press, London.

Grantham-McGregor, S., Cheung, Y. B., Cueto, S., Glewwe, P., Richter, L., & Strupp, B. (2007). Developmental potential in the first 5 years for children in developing countries. *The Lancet, 369*(9555), 60–70.

Kaffenberger, M., Silberstein, J., & Spivack, M. (2022). *Evaluating systems: Three approaches for analyzing education systems and informing action, RISE W. P. Series 22/093*. https://doi.org/10.35489/BSG-RISE-WP_2022/093

Koc, N., & Çelik, B. (2015). The impact of number of students per teacher on student achievement. *Procedia - Social and Behavioral Sciences, 177*, 65–70.

Kramarz, F., Nimier-David, E., & Delemotte, Th. (2022). Inequality and earnings dynamics in France: National policies and local consequences. *Quantitative Economics, 13*, 1527–1591. 1759-7331/20221527

Locatelli, R. (2019). *Reframing education as a public and common good*. Palgrave Macmillan. https://doi.org/10.1007/978-3-030-24801-7_

Msezane, S. (2020). School evaluation: Approaches, frameworks, and indicators. *Quality Education*, 1–12. https://doi.org/10.1007/978-3-319-69902-8_8-1

Niemi, H., & Jakku-Sihvonen, R. (2011). Teacher education in Finland. In M. Valenčič Zuljan, J. Vogrinc (Eds.), *European dimensions of teacher education. Similarities and differences* (pp. 33–52). Ljubljana Faculty of Education and Kranj The National School of Leadership and Education.

Nieuwelink, H., Dekker, P., & ten Dam, G. (2019). Compensating or reproducing? Students from different educational tracks and the role of school in experiencing democratic citizenship. *Cambridge Journal of Education, 49*(3), 275–292.

OECD. (2011). *Education at a glance, OECD indicators 2011*. OECD Publishing. https://www.OECD.org/education/skills-beyond-school/48631582.pdf

OECD. (2012). *Education at a glance, OECD indicators 2012*. OECD Publishing. https://www.OECD-ilibrary.org/education/education-at-a-glance-2012_eag-2006-en

OECD. (2016). *PISA 2015 results (volume II): Policies and practices for successful schools*. PISA, OECD Publishing. https://doi.org/10.1787/9789264267510-en

OECD. (2017). *PISA in focus 2017/76, September*.

OECD. (2019). *Education at a glance*. https://www.OECD-ilibrary.org/education/education-at-a-glance-2019_f8d7880d-en

OECD. (2020). *PISA 2018 results (volume III), What school life means for students' lives*. https://www.OECD-ilibrary.org/sites/f05bb3ee-en/index.html?itemId=/content/component/f05bb3ee-en

OECD. (2023). *Education GPS*. https://gpseducation.oecd.org/revieweducationpolicies/#!node=44129&filter=all

Pedró, F., Leroux, G., & Watanabe, M. (2015). *The privatization of education in developing countries: Evidence and policy implications*. UNESCO. https://hdl.handle.net/20.500.12799/4013

Pensiero, N., Giancola, O., & Barone, C. (2019). Socioeconomic inequality and student outcomes in Italy. In: L. Volante, S. Schnepf, J. Jerrim, & D. Klinger (Eds.), *Socioeconomic inequality and student outcomes. Education policy & social inequality* (Vol. 4). Springer. https://doi.org/10.1007/978-981-13-9863-6_5

Philip, G., De Wit, H., & Ayenachew, A. (2021). *Public vs. private participation in higher education: Realities and debates*. UNESCO, ED/GEMR/MRT/P1/11

Popova, M. (2015). *An ideal education system*, Sep 19. https://medium.com/@mariannapopova/an-ideal-education-system-1a5cb73d5b59

Redaelli, S. (2020). The obsolescence of the current Italian compulsory education system. *Society & Culture*, June 20. https://culturico.com/2020/06/20/the-obsolescence-of-the-italian-education-system/

Riordan, S. (2022). Improving teaching quality to compensate for socio-economic disadvantages: A study of research dissemination across secondary schools in England. *Review of Education, 10*(2), 1–34.

Sengeh D., & Winthrop, R. (2022). *Transforming education systems: Why, what, and how*, 23 June. https://www.brookings.edu/articles/transforming-education-systems

Shonkoff, J. P., & Garner, A. S. (2012). The lifelong effects of early childhood adversity and toxic stress. *Pediatrics, 129*(1).

Suárez-Orozco, C., Yoshikawa, H., & Tseng, V. (2015a). Intersecting inequalities: Research to reduce inequality for immigrant-origin children and youth. *Sociology, Education*, 1 February.

Suárez-Orozco, C., Yoshikawa, H., & Tseng, V. (2015b). *Intersecting inequalities: Research to reduce inequality for immigrant-origin children and youth.* William T. Grant Foundation Paper.

Verger A., Zancajo, A., & Fontdevila, C. (2022). *Private sector participation in education: where it occurs, and why*, 24 February 2020 (updated on 1 February 2022)

Wood-Wallace, D. (2016). *The role of assessment in teaching and learning.* Academia.

World Economic Forum. (2021). *The Davos Agenda 2021. What will education look like in 20 years?* Jan 28. https://www.weforum.org/agenda/2021/01/future-of-education-4-scenarios/

Uncited References

Bhattacharjee, M. (2021). What is the role of a teacher in a student's life? *Edumilestones Blog.* https://www.edumilestones.com/blog/details/Role-of-a-Teacher-in-a-Students-Life

Beach, D. (2018). The myth of Swedish education equity. In D. Beach (Ed.), *Structural injustices in Swedish education.* Palgrave Macmillan, Cham, https://doi.org/10.1007/978-3-319-95405-9_1, pp. 1-32

Bisht, S. (2022). *Teaching models: All you need to know*, April 14. https://edusights.com/teaching-models/

Britannica. https://www.britannica.com/topic/education/Education-in-classical-cultures

Britannica, Education—Medieval, Monastic, Literacy. https://www.britannica.com/topic/education/Europe-in-the-Middle-Ages

Cullinan, J., & Cawley, J. (2017). Parental misclassification of child overweight/obese status: The role of parental education and parental weight status. *Economics & Human Biology, 24*, 92–103.

Dewey, J. (1974). *John Dewey on education: Selected writings.* University of Chicago Press, Chicago

Erasmus+ *EU programme for education, training, youth and sport.* https://erasmus-plus.ec.europa.eu/programme-guide/part-a/priorities-of-the-erasmus-programme/objectives-features

Expatica. (2023). *The education system in Portugal.* https://www.expatica.com/pt/education/children-education/the-education-system-in-portugal-105195/

European Commission. (2022d). *Erasmus+ Programme Guide*, Version 1 (2022), 24-11-2021.

European Commission, European Education Area. *Quality education and training for all.* https://education.ec.europa.eu/education-levels/higher-education/relevant-and-high-quality-higher-education

European Commission (no date). *European education area. Quality education and training for all.* https://education.ec.europa.eu/about-eea/the-eea-explained

European Commission, EACEA, Eurydice (2022a). *The structure of the European education systems, 2022/2023. Eurydice Facts and Figures.* Publications Office of the European Union.

European Commission, EACEA, Eurydice (2022b). *The Organisation of school time in Europe. Primary and general secondary education—2022/2023.* Eurydice Facts and Figures. Publications Office of the European Union.

European Parliament (2021). *The European Education Area and the 2030 strategic framework for education and training.* https://www.europarl.europa.eu/RegData/etudes/BRIE/2021/690630/EPRS_BRI(2021)690630_EN.pdf

Eurostat. *Statistics explained.* https://ec.europa.eu/eurostat/statistics-explained/index.php?title=Main_Page
Eurostat. *Young people on the labour market.* Statistics. https://ec.europa.eu/eurostat/statistics-explained/index.php?title=Young_people_on_the_labour_market_-_statistics
Frønes, T. S., Pettersen, A., Radišić, J., & Buchholtz, N. (2020). Equity, equality and diversity in the Nordic model of education. Contributions from large-scale studies. In T. S. Frønes, A. Pettersen, J. Radišić, & N. Buchholtz (eds.), Equity, equality and diversity in the Nordic model of education. Springer. https://doi.org/10.1007/978-3-030-61648-9_1
Hanushek, E. A., & Woessmann, L. (2015). *The knowledge capital of nations: Education and the economics of growth.* MIT Press, Cambridge, Mass.
Harrington-Atkinson, T. (2021). *The Learning Pyramid*, February 28. https://tracyharringtonatkinson.com/the-learning-pyramid/
Henrekson, M., & Wennström, J. (2022). Educational performance in Swedish schools—What are the facts? In M. Henrekson & J. Wennström (Eds.), *Dumbing down. The crisis of quality and equity in a once-great school system—and how to reverse the trend.* Palgrave Macmillan.
Moos, L., Möller, J., & Johansson, O. (2004). A Scandinavian perspective on educational leadership. *Educational Forum, 68*(3), 200–210
OECD. (2023). *PISA 2022 results (volume II): Where all students can succeed.* OECD Publishing. https://www.OECD.org/publications/pisa-2018-results-volume-ii-b5fd1b8f-en.htm
OECD. Evaluation and assessment frameworks for improving school outcomes. *Common Policy Challenges.* www.OECD.org/edu/evaluationpolicy
Our World in Data. https://ourworldindata.org/global-education
Our World in Data. https://ourworldindata.org/grapher/out-of-school-children-of-primary-school-age-by-world-region
Our World in Data. https://ourworldindata.org/grapher/gross-vs-net-enrollment-rate-in-primary-education?country=~MDG
Statista. https://www.statista.com/chart/11058/bachelor-tuition-fees-international-comparison/
Telhaug, A. O., Mediås, O. A., & Aasen, P. (2006). The Nordic model in education: Education as part of the political system in the last 50 years. *Scandinavian Journal of Educational Research, 50*(3), 245–283. https://doi.org/10.1080/00313830600743274
UNESCO. https://data.worldbank.org/indicator/SE.TER.ENRR.FE?locations=ZJ)
UNESCO. *Priority gender equality.* https://www.unesco.org/en/gender-equality/education
Whitty, G. (2001). Education, social class and social exclusion. *Journal of Education Policy, 16*(4), 287–295
Wikipedia. *Education.* https://en.wikipedia.org/wiki/Education
Wikipedia. *Education in Africa.* https://en.wikipedia.org/wiki/Education_in_Africa
Wikipedia. *Education in Asia.* https://en.wikipedia.org/wiki/Education_in_Asia
Wikipedia. *Education in Australia.* https://en.wikipedia.org/wiki/Education_in_Australia
Wikipedia. *Education in Canada.* https://en.wikipedia.org/wiki/Education_in_Canada
Wikipedia. *Education in Latin America.* https://en.wikipedia.org/wiki/Education_in_Latin_America
Wikipedia. *Education in New Zealand.* https://en.wikipedia.org/wiki/Education_in_New_Zealand
Wikipedia. *Education in the United States.* https://en.wikipedia.org/wiki/Education_in_the_United_States
Wikipedia. *Teaching method.* https://en.wikipedia.org/wiki/Teaching_method
World Bank. https://wdi.worldbank.org/table/2.7
World Bank. (2023). *Education*, April 11. https://www.worldbank.org/en/topic/education/overview

SPRINGER NATURE

GPSR Compliance

The European Union's (EU) General Product Safety Regulation (GPSR) is a set of rules that requires consumer products to be safe and our obligations to ensure this.

If you have any concerns about our products, you can contact us on ProductSafety@springernature.com

In case Publisher is established outside the EU, the EU authorized representative is:

Springer Nature Customer Service Center GmbH
Europaplatz 3
69115 Heidelberg, Germany

The manufacturer's authorised representative in the EU is Springer Nature Customer Service Centre GmbH, Europaplatz 3, 69115 Heidelberg, Germany. If you have any concerns regarding our products, please contact ProductSafety@springernature.com

Printed and bound by CPI Group (UK) Ltd, Croydon, CR0 4YY
01/10/2026
02225985-0003